VISIONS OF THE UNIVERSE

MITCHELL BEAZLEY

VISIONS OF THE UNIVERSE

THE LATEST DISCOVERIES IN SPACE REVEALED

DR RAMAN PRINJA

Visions of the Universe
by Dr Raman Prinja

First published in Great Britain in 2004 by Mitchell Beazley,
an imprint of Octopus Publishing Group Limited,
2–4 Heron Quays, London E14 4JP.

© Octopus Publishing Group Limited 2004
Text © Raman Prinja 2004

All rights reserved. No part of this publication may be
reproduced or utilized in any form by any means, electronic or
mechanical, including photocopying, recording, or by any
information storage and retrieval system, without prior written
permission of the publisher.

ISBN 1 84000 974 8

A CIP catalogue record for this book is available from the
British Library

Typeset in Univers
Printed and bound in China

*In loving dedication to Kamini, Vikas, and Sachin, and my parents
and family.*

While all reasonable care has been taken during the preparation
of this edition, neither the publisher, editors, nor the author can
accept responsibility for any consequences arising from the use
thereof or from the information contained therein.

Commissioning Editor Vivien Antwi
Editorial Assistants Sophie Wardell, Bryn Cross
Design Peter Gerrish
Production Gary Hayes
Picture Research Giulia Hetherington
Copy Editor Andrew Hilliard
Proofreader Siobhan O'Connor
Indexer Sandra Shotter

Opposite page: This detail of an image, taken from the *Cassini*
spacecraft, is one of the most detailed global images ever taken of
Jupiter. The reddish-brown bands and white ovals of the dynamic
atmosphere are clearly visible.

Contents

Introduction

The exciting new astronomical discoveries and fascinating advances that are now taking place in our knowledge of the universe are brought to light in this book, through some of the most magnificent images ever obtained. The study of space is flourishing today thanks to giant telescopes perched on mountain-tops and powerful observatories in orbit around Earth, such as the Hubble Space Telescope. The exquisite views from these remarkable instruments are complemented by high-resolution images beamed back by the latest spacecraft missions to Mars, Jupiter, and our Moon. In *Visions of the Universe*, we take a look at how these revealing observations are expanding the limits of our understanding, and the answers they provide to some of the most fundamental questions of the universe.

Our voyage through space in this book spans an enormous variety of phenomena. Starting from spectacular recent evidence of liquid water on Mars and the moons of Jupiter, we progress to giant clouds of gas that host embryonic stars, then beyond to the grandest structures in the universe assembled from millions of galaxies. We also witness the most powerful and violent events in the universe, marked by the latest visions of exploding stars and awesome collisions between galaxies. The high-energy processes also extend to energetic convulsions of ageing stars and cauldrons of intensely hot, swirling gas that surround voracious black holes. The quest to understand an astonishing universe drives us toward the most perplexing conundrums facing astronomers today. These include the mysterious nature of dark matter and the even more enigmatic dark energy that may be acting to push the expansion of the universe into overdrive.

It is useful to start with a brief guided tour to review our modern perspective of the main levels of structure in the universe, beginning nearby in the solar system and progressing to the grandest scales in space. Our planet Earth is one of nine that belong to the solar system, which also consists of almost a hundred moons so far discovered, plus innumerable chunks of ice and rocks in the form of comets and asteroids. The Sun is the dominant member of the solar system, providing more than 99 per cent of the total mass and most of the energy to heat the planets.

The Sun is an ordinary star, like many of the stars in the night sky. The thousands of stars we can see with our naked eyes are, however, just a tiny fraction of the 100 billion stars that collectively make up the spiral-shaped structure that is our Milky Way Galaxy. Our solar system has a rather mediocre location in the Galaxy, well away from the Galaxy's centre. Galaxies are great islands of stars that come in a range of shapes, from magnificent whirl-like spirals to vast elliptical or egg-like structures. Some galaxies have no discernible shape, often because they have undergone major disruption due to collisions with other galaxies. The majority of galaxies congregate into clusters, comprising

The spiral-shaped galaxy called NGC300, as viewed through a telescope at the European Southern Observatory (ESO) at La Silla, Chile. Located about seven million light years from Earth, NGC300 contain billions of stars and is akin to the Milky Way Galaxy. Grand galaxies such as this one are the building blocks that make up the large-scale structure of the universe.

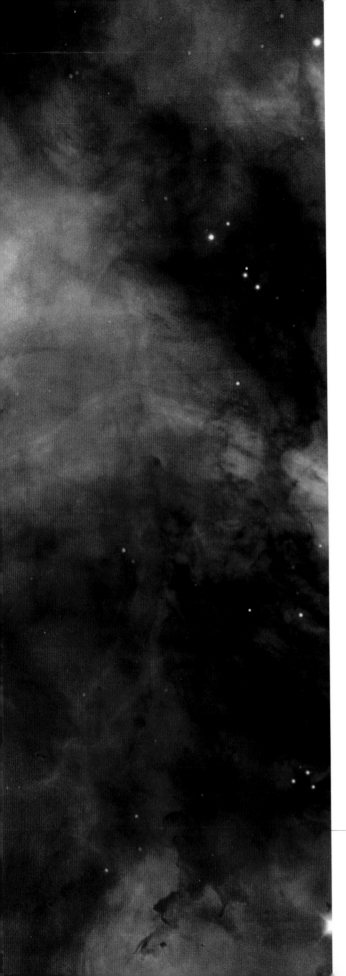

dozens or even hundreds of members held together by their mutual gravity. Clusters of galaxies start to define the large-scale design of the universe. On the grandest scales, the more than 100 billion galaxies in the observable universe are arranged in vast chains and sheets, which are spaced by huge voids or "holes" that contain very few galaxies. One of the great pursuits of modern cosmology is to understand the origin of this large-scale design, which gives the universe a "spongy" or "froth-like" appearance.

A major triumph of 20th-century astronomy was the discovery of evidence that the entire universe is expanding and everything must have been closer together in the past. Turning the "clock" back, the universe passed through an incredibly hot and dense phase in the distant past, which relates to its origin about 14 billion years ago, referred to as the "Big Bang". The universe has continued to expand ever since the Big Bang, though of course gravity has succeeded over comparatively small scales to gather matter to create structures such as the stars and galaxies. The Big Bang picture has today become our framework for understanding the origin, evolution, and fate of the universe.

The vastness of space

The universe is incredibly vast and its description involves astonishing sizes, distances, and ages that are beyond our normal perspectives on Earth. Light travels in space at a remarkable 300,000 km per second (186,000 miles per second), but this is a finite speed which means that even light can take a substantial time to span the enormous distances in the universe. Light from the Moon takes about a second to reach us, and about eight minutes to traverse the 150 million km (93 million miles) from the Sun to the Earth. Light from the nearest stars to the Sun takes many years to reach us. The distance that light travels in one year is called a "light year", and is a fundamental unit of length in astronomy; it is equal to almost 10,000 billion km (written as 1 followed by 13 zeros; 6000 billion miles), which is already an unfamiliar distance in our everyday lives, but trivial in the expanse of space. Our Galaxy, with

This image from NASA's Hubble Space Telescope (HST) shows intricate details within a vast star-forming region called the Carina Nebula. Also embedded in the clouds of gas and dust are some newborn stars that are considerably more powerful than the Sun.

its swarm of 100 billion stars, has a diameter of 100,000 light years. The next nearest grand spiral galaxy to ours, the Andromeda, is about 2.5 million light years from Earth. This means that light from it takes around 2.5 million years to travel to us!

As light signals take time to travel and do not reach us immediately, the further away we look into the universe, the further back we are seeing in time. This remarkable fact allows us to use powerful telescopes and look at regions of the universe as they appeared in the past. The universe is so enormous that when we receive light from the furthest galaxies, it has been travelling for billions of years. We are seeing these remote galaxies as they were billions of years ago, in their youth. To understand how the galaxies have changed since that time, and what they have evolved into, astronomers carry out extensive observations of a variety of other galaxies across a range of distances (and therefore ages). By building a detailed picture of the cosmic "zoo", it is possible to compare the most distant galaxies (viewed when they are newly formed) to relatively nearby ones (which have gracefully aged). Acting like detectives, astronomers can then piece together the probable history of galaxies.

For more than 300 years, up to the second half of the 20th century, astronomical observations were carried out using optical telescopes, which detect only the visible components of light. That makes sense, of course, as our eyes are sensitive only to visible light. Through optical telescopes we have come to develop our most familiar picture of the universe, defined by its main constituents of gas, stars, and galaxies. There are, however, other types of radiation such as radio waves, ultraviolet waves, and X-rays. These non-visible regions of the electromagnetic

spectrum are also highly valuable windows on the universe. The problem for astronomers is that the Earth's atmosphere is very effective in blocking most radiation other than visible light and radio waves. To overcome the shield provided by the atmosphere new types of telescopes have been developed and placed in satellites that orbit above the Earth's atmosphere. These are exciting and advanced facilities that offer entirely different views of the universe, spanning infrared to X-ray wavebands. The hottest gas and most energetic objects in space are revealed by telescopes that can detect gamma-rays and X-rays, including exploding stars and turbulent regions around massive black holes. Observations are made in the infrared wave-band to study the cool universe. In our Galaxy the cool picture is dominated by molecules and microscopic dust particles loaded into giant clouds called "nebulae" that are the star-making factories of space.

The latest revelations in this book represent an outstanding combination of discovery, exploration, and innovation. The beauty of the images surveyed in the following chapters casts new light on the universe. We look at how the views are interpreted, and evaluate their impact on our quest to understand the origin and nature of the key components of the universe.

The European Space Agency (ESA)'s Mars Express mission captured this detailed view on 15 January 2004, from a height of 273km (170 miles) above the Martian surface. The 100km (60-mile) wide scene shows a channel formed billions of years ago by water that once flowed there. The hunt for liquid water and primitive life forms is one of the major drivers in our current exploration of the solar system.

Searching Mars' watery past: NASA's Columbia Memorial Station sits on the floor of Gusev crater on Mars. The *Spirit* rover was released from the station during January 2004 to search for evidence of ancient environments that were once flooded with water.

Follow the water

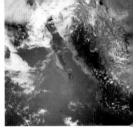

Water is an essential prerequisite for almost every known form of life on Earth. Accordingly, scientists are eagerly searching for its liquid form in the solar system beyond our planet. If discovered, such locations would offer the most intriguing clues as to whether primitive life exists elsewhere. Recently obtained data raises the tantalizing possibility that liquid water may be present below the dry surface of Mars and beneath the frozen upper layers of Europa, one of the larger moons of the giant gas planet Jupiter. There is also intriguing new evidence pointing to the existence of frozen water in craters at the polar regions of Earth's Moon.

So why do scientists, including astronomers, planetary geologists, and astro-biologists, "follow the water"? Water is an essential component of life on Earth. Almost every location of liquid water here has been known to support life. More than two-thirds of the Earth's surface is covered in water, 97 per cent of which is found in vast oceans, with the Pacific Ocean alone covering almost 170 million square km (65 million square miles) .

Life generally requires liquid water to form and survive. It is an excellent solvent for biochemistry, providing a medium where extensive chemical reactions can occur. Complex organic molecules can develop, including those found in the cells of living organisms. Water remains in liquid form over a wide range of temperatures, which can be of great benefit when life is faced with severe weather and climate changes.

This is one of the most detailed true-colour images ever obtained of the Earth. The view is based on a collection of satellite observations, and it is essentially a mosaic of every square kilometre of our planet. The Earth is the only planet in the solar system with vast oceans of liquid water on its surface and an abundance of life. Typically, at any given time, more than half our planet is also covered by water-bearing clouds.

A wet Mars

For the past two centuries there has been a fascination with Mars and the prospect that life may exist there. Our understanding of Mars has changed dramatically over the past 30 years. During the 1970s, NASA's spacecraft *Mariner 9*, *Viking 1*, and *Viking 2* orbited the Red Planet and sent back images revealing details of valleys, channels, and canyons that were formed by water flowing billions of years ago. Incredibly heavy water flows, equivalent to thousands of Mississippi rivers, would have been needed to shape some of the features seen on the surface of Mars. For example, the vast system of canyons known as Valles Marineris extends for 4000km (2500 miles), and some parts are over 7km (4 miles) deep and 600km (370 miles) wide. This makes it four times longer, and five times deeper, than the Grand Canyon in the United States.

Ancient water

Since 1999 NASA's spacecraft *Mars Global Surveyor* has returned remarkably detailed images of sediment layers on Mars. Sediments on Earth are layers of rocky material that have been deposited by the motion of large amounts of water. The features on Mars support the notion of stable flowing water early in the planet's history. The observations highlight a layered texture, which suggests that material has been deposited in a repeated fashion in a lake or shallow sea. One example is the layered rock on a region called Candor Chasma, which is part of the Valles Marineris canyons. More than a hundred rock-bed layers are seen in this area, each about 10m (11 yards) thick, with a smooth upper surface and steep cliffs at the edges. One possibility is that the layers were built up by a dynamical under-water environment. An example of this on Earth is the erosion enacted by the Colorado River, which has exposed the sediment layers that can be seen today in the walls of the Grand Canyon. These interpretations of sediment layers on Mars indicate that substantial regions, which were once dynamical underwater environments, may in the past have been capable of supporting life.

Sediment layers have also built up from material that settled on the floors of craters that were likely once filled with water. Billions of years ago the bottoms of these "ponds" were eroded to reveal successive layers of sedimentary rock. One such site is Gusev crater, which is almost 170km (105 miles) across and was created by the impact of an asteroid. In the subsequent past it

appears to have been flooded by water flowing through a huge valley called Ma'adim Vallis. It is likely that the Gusev crater remained wet for long periods, making it an ideal place to search for fossil evidence of primitive life forms. Indeed, this was the landing site chosen for one of the two robotic exploration rovers successfully placed on Mars by NASA during January 2004. Named *Spirit*, the rover has now begun its mission to determine the history of climate and water in Gusev crater. Using a sophisticated set of instruments, *Spirit* is being manoeuvred from Earth to probe whether selected sites were once wet enough to support life. The second rover, named *Opportunity*, landed in a region called Meridiani Planum, about halfway round the planet from Gusev crater. Both explorers will travel around their landing sites to study rocks and soil using a set of five geology instruments, plus a special abrasion tool that can be used to scrape the rocks to expose fresh surfaces. The primary goal of both *Spirit* and *Opportunity* is to evaluate whether the conditions on Mars may once have been favourable to life.

Channels caused by the action of running water that flowed billions of years ago are also common features on Mars. Reminiscent of rivers draining on Earth, their formation is thought to be due to two main mechanisms: first, as a result of runoff after

A spectacular panoramic view of the surface of Mars captured by the cameras on NASA's *Spirit* exploration rover, which landed on the Red Planet on 3 January 2004. The peaks of the hills on the horizon are 3km (2 miles) from the robotic rover.

rainfall, which carved into the rocky surrounding material, and secondly as a result of underground water emerging from the collapse of surface layers. Many of the channels date from periods of more than three billion years ago. Large, long-extinct volcanoes on Mars also reveal water-carved channels on their slopes, pointing to a link between ancient volcanic heating and large-scale floods. The volcanic activity may have melted large volumes of underground water ice, causing massive flows that subsequently formed deep channels and networks of valleys. Subsurface reservoirs of liquid water, warmed by nearby volcanoes, would also have intriguing potential as a habitat for primitive Martian life forms.

Water has obviously affected the geological history of Mars, but it flowed a long time ago; today, Mars appears predominantly as a dry and dusty desert. The rivers, and possibly even oceans,

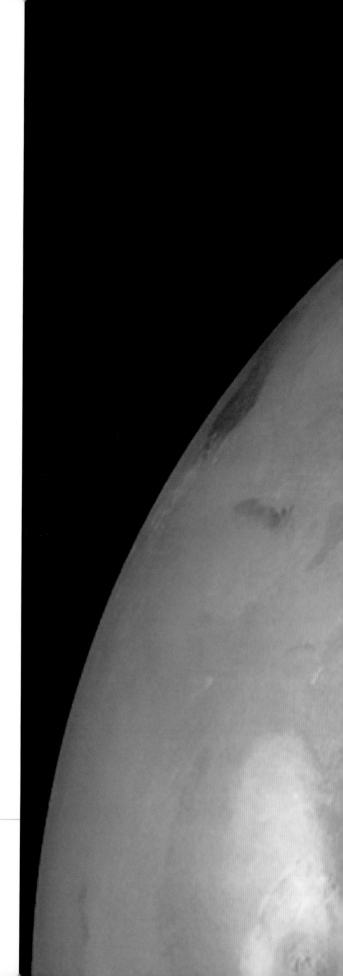

have long since dried up, boiled away billions of years ago by intense solar radiation. Orbiting the Sun at a distance one and a half times greater than the Earth, Mars intercepts less than 50 per cent of the solar energy received by our planet; it also has a gravity of about one third of Earth's. Consequently, its carbon dioxide atmosphere traps little heat and is too thin and cold for water to remain liquid at the surface. Despite this, scientists predict that some water should still be present underground today, shielded from the surface conditions above. Dramatic new evidence for this water has come to light over the past few years, in the form of exquisite images taken by the latest spacecraft missions sent to study the planet.

Fresh water, too

The most recent discoveries about Mars are also some of the most dramatic. During 2000 and 2001 NASA's *Mars Global Surveyor* spacecraft revealed remarkable gullies on the surface of the planet which are thought to have been carved relatively recently by the drainage of liquid from the surface. The gullies appear on images of some craters, troughs, and valleys. Compared to the 4.5-billion-year age of the planet, they are geologically young, perhaps less than a few million years old. Furthermore, in some cases, the gullies could even be seeping water today. Evidence of the gullies' youth comes from the crispness of their edges, and the fact that they sometimes lie on top of wind-blown regions that are themselves recent. The gullies are so relatively young that they have raised new debate on the possible presence of liquid water on Mars today. Certainly, if bacterial life has developed on Mars, it's quite probable it would have done so in these locations.

The gullies on the surface are typically a kilometre (0.6 miles) long and half a kilometre (0.3 miles) wide. They look like trenches created by flowing water, often showing deposits of rock and soil that have been carried by these flows. Examples of these features on Earth are the debris flows seen in Greenland and Canada, which usually occur over permafrost. Scientists have

Winter in the southern hemisphere of Mars imaged in late 2001 by the *Mars Global Surveyor* spacecraft. The circular patches in the lower left of the image are clouds of water ice, which are seen hovering over long-extinct volcanoes. Some water ice is also present in the summertime north polar cap at the top of the image.

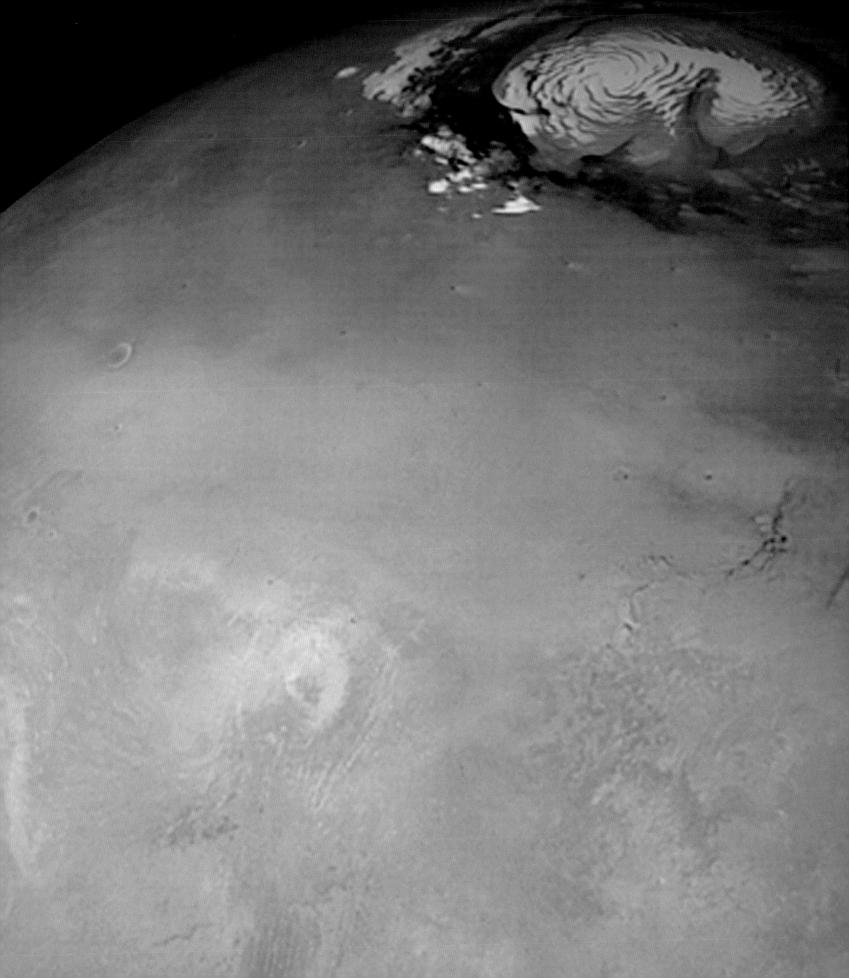

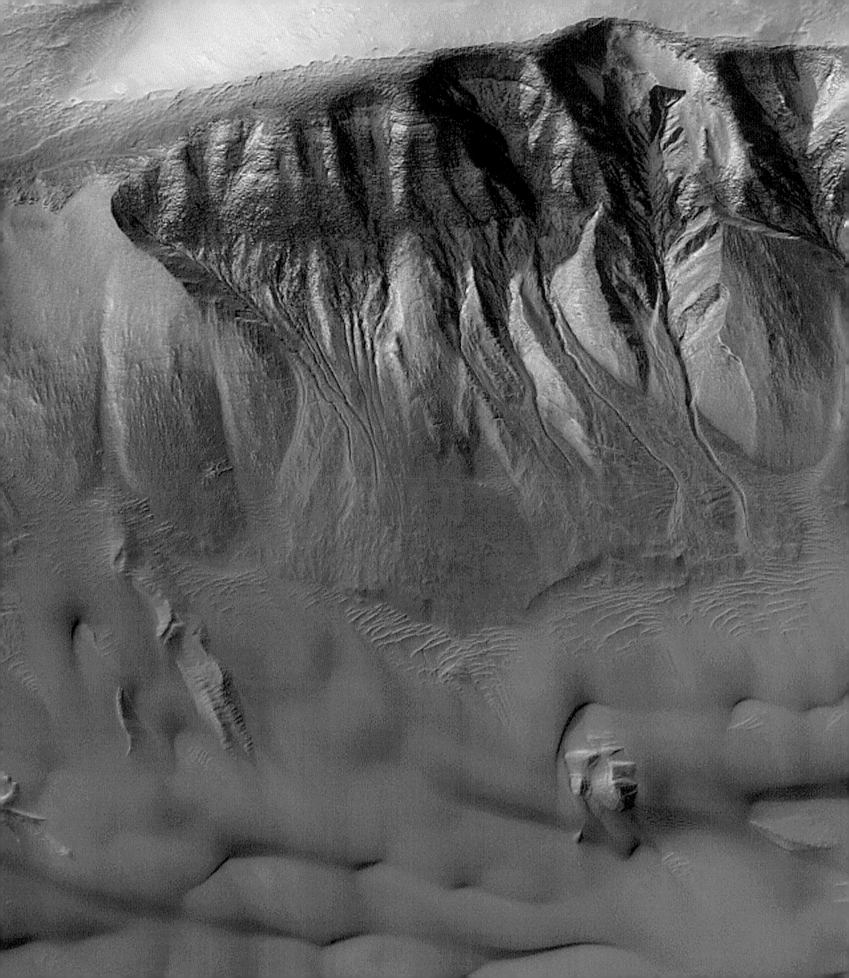

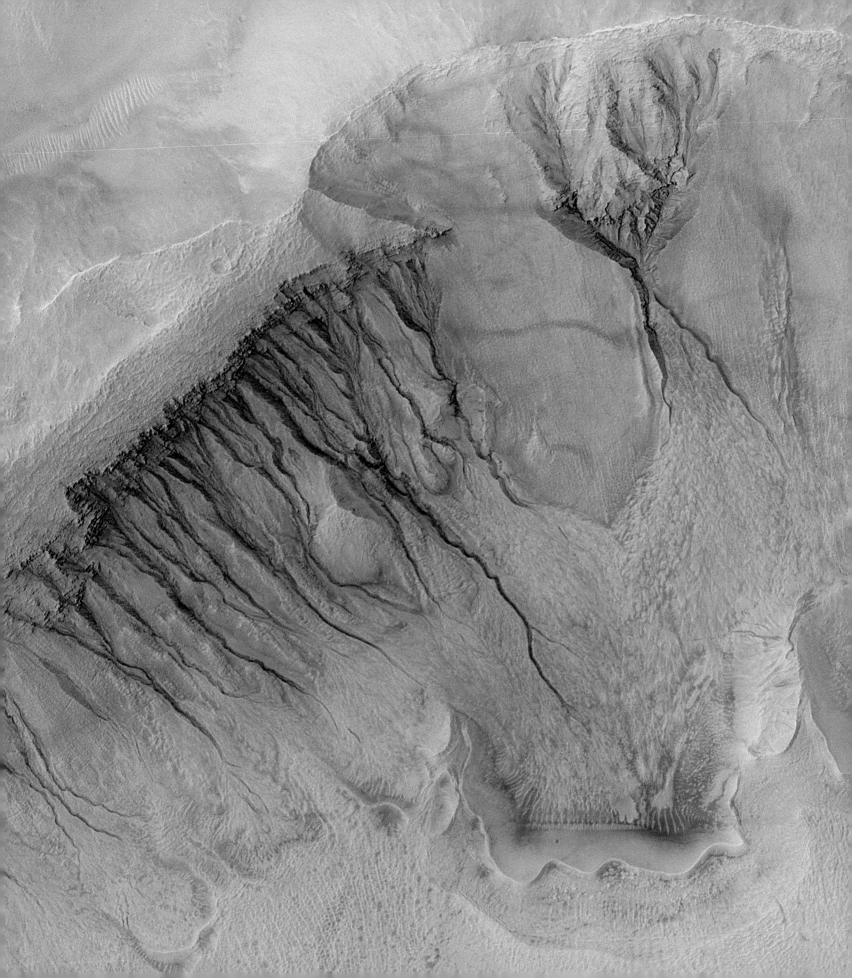

Previous page: These two images taken by NASA's *Mars Global Surveyor* reveal dramatic evidence for gullies on the walls of two different craters. The gullies appear to have formed when subsurface water seeped out to the Martian surface.

estimated that a volume equivalent to several large swimming pools would need to have flowed in order to form the ditches photographed on Mars.

The exact process by which these gullies on Mars were created is unclear. Due to its thin atmosphere, the pressure at the planet's surface is nearly a hundred times less than at sea level on Earth, and liquid water would immediately boil if exposed above the ground. It is thus thought that the water supply is located 100 to 300m (110 to 330 yards) below the Martian surface, locked up in the form of ice, and probably confined to small regions across the planet. Volcanic heating and geothermal activity help to create outbursts of water, similar to flash floods on Earth. Frozen water behind an initial seepage site gradually collects into an ice dam, and the pressure builds until the dam breaks to push a flood down a gully.

The exciting possibility raised by the discovery of these gullies is that liquid water may be present today in certain underground areas of Mars. If water and energy promote life on Earth, then there are reasons to expect the same may be true on Mars. This possibility has significant implications for human exploration of the planet. If water were present in substantial amounts in regions other than the north and south poles of Mars, it would be easier to access and exploit it. Water would be a crucial resource for human exploration of Mars. Aside from its use for drinking, oxygen can be extracted from it to create breathable air, and oxygen can also be extracted and combined with hydrogen to create fuel cells to provide electrical power. There is no doubt that human exploration of Mars will soon enter an exciting new era.

The search for water (and possible primitive life) in the solar system has been buoyed by recent discoveries that life on Earth is present in conditions of extreme heat, cold, acidity, and radiation. It is remarkable that bacteria thrive in the bubbling hot springs of Yellowstone National Park in the United States, which are highly acidic and reach temperatures of almost 90°C (195°F). By contrast, communities of microbes thrive on Earth at temperatures as low as -18°C (0°F). Even in the driest and hottest deserts on Earth, such as Atacama in the Andean planes of Chile, life exists on minimum moisture in the form of lichens and fungi. Harsh and extreme environments comparable to these could quite possibly exist on other planets and moons in our solar system.

While the advantages of water as a medium for supporting life on Earth are beyond question, it is unknown whether other liquids such as ammonia or methane could also promote and support it. This scenario has given rise to numerous sci-fi tales of exotic life forms in the most hostile locations in space. The reality is that the quest for evidence of life in our solar system is based on our understanding of the role of liquid water; it remains a prime medium for a habitable world, and an abundant one in our solar system. This is the foundation for some of the most exciting exploration and research of the solar system today.

Surprising layers of sedimentary rock on the surface of Mars are revealed here in remarkable detail. The 1.5km by 3km (1 mile by 1.9 miles) region shown contains more than 100 beds, with smooth surfaces and steep boundary cliffs. The patterns may be due to material deposited billions of years ago in a lake or shallow sea.

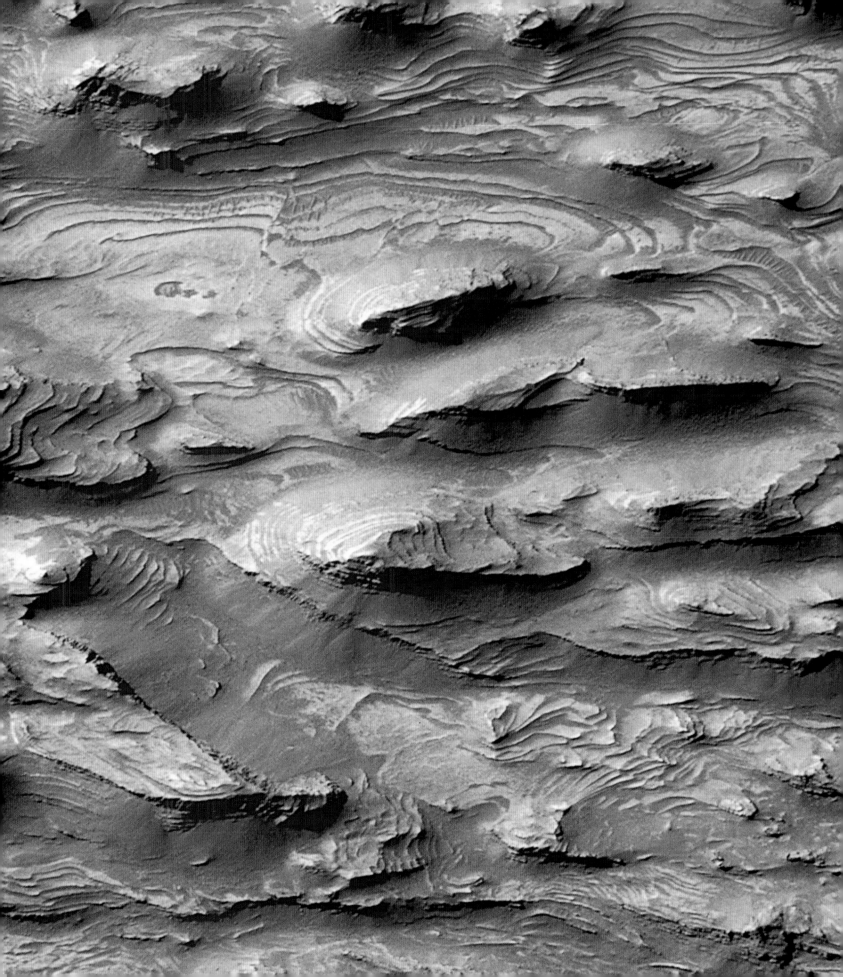

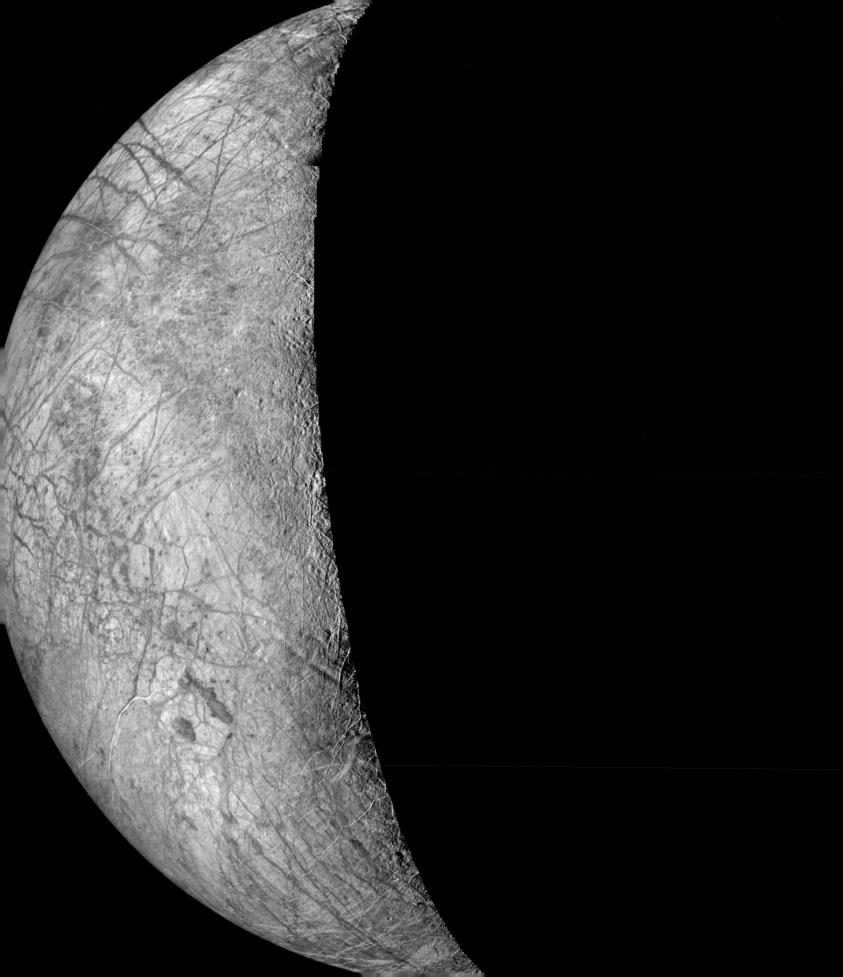

Oceans on Europa

A distant moon in the outer regions of the solar system, lying almost 800 million km (500 million miles) from the Sun, would seem a most unlikely location for liquid water. Remarkably, though, between 1997 and 2002 NASA's *Galileo Orbiter* spacecraft uncovered exciting evidence for a fermenting ocean of water beneath the icy surface of Europa, Jupiter's fourth-largest moon.

Europa is a little smaller than Earth's Moon, and is one of the four originally discovered orbiting Jupiter by Galileo Galilei in 1610. The brightest of Jupiter's moons, it has a markedly different appearance from that of almost every other planet or moon in the solar system, its ice-coated surface giving it a 70 per cent reflectivity. This ice coating is very pure, suggesting that it is replenished from fresh material below the surface. Europa is also the smoothest body of this class in the solar system, with very few impact craters and no mountains on its surface. There is no doubt, however, that it would have been bombarded by numerous comets and asteroids billions of years ago, in the early history of the solar system. The fact that evidence for these impacts is mostly erased today indicates that Europa's current surface is

both resilient, in that it can absorb the shock from an asteroid without forming a vast crater, and very young, with impacts being repaired by fresh coatings of ice from below the surface. By comparison, Mars today has several hundred craters larger than 100km (60 miles) in diameter, and numerous smaller ones.

The *Galileo* spacecraft mission has provided us with the closest views of Europa yet seen, with some images taken at less than 600km (370 miles) above its surface. A bewildering variety of features was revealed, which, combined with other measurements and readings, provides intriguing details of substantial subsurface oceans of salt water. The presence of these oceans and their implications are discussed below.

Ironically, these exciting discoveries about Europa determined the ultimate fate of the spacecraft itself. To avoid any possibility of contaminating the pristine moon, when *Galileo* reached the end of its highly successful mission in September 2003, it was deliberately plunged into Jupiter's dense atmosphere, to destroy it.

Floating ice rafts

Unique to Europa are detailed images of blocks of ice that appear to have broken apart and rearranged themselves, as though they were floating. The same chaotic pattern of fractured and rotated blocks can be seen in ice floes in the Arctic regions of the Earth. The individual blocks of ice on Europa may be several kilometres across and at least several hundred metres deep, and some are tipped or partially submerged, as may be expected of icebergs floating in a liquid water ocean.

A beautiful crescent view of Jupiter's moon Europa. The patchy spread of surface ice is set against long linear fractures that cross the surface of the moon in various directions.

Marked plains

The most common surface types on Europa are smooth and ridged plains. The former could be the result of a kind of water-ice volcanism, where liquid erupts from below the surface, spreads over a wide region, then freezes. The ridges could have been formed when some of the ejecta was thrown up as fluid, and may have settled as piles of debris. A typical ridge may be a few hundred metres high, often with a groove or trench running along its centre. Their appearance is consistent with the notion of fractures at the surface subsequently filled by water eruptions from below. Ridges similar to these have been studied in ice packs in the Arctic on Earth. There are also numerous long and dark lines that mark the surface of Europa's plains, extending for thousands of kilometres. Some of the most notable markings outline curved or scalloped patterns which are thought to have formed very rapidly, at least from a geological viewpoint. The curved and intricate bands signal fractures that have broken the icy crust of Europa into plates, which may be lubricated by soft ice or liquid water below. Subsurface liquid also rises to subsequently fill the gaps from splits in the icy crust.

Magnetic readings

Some of the strongest evidence for the existence of a subsurface ocean on Europa comes from magnetic readings of the moon taken by the *Galileo* spacecraft. The giant gas planet Jupiter has an intense magnetic field that extends well beyond the orbits of its numerous moons, forming a vast region called a "magneto-

sphere". During the *Galileo* mission measurements were made of Europa as it travelled within this magnetosphere. The magnetic field of Jupiter was discovered to set up electric currents in the upper 100km (60 miles) of Europa's surface. This experiment showed that there was a conducting layer present. Furthermore, systematic changes in the properties of these magnetic readings were consistent with the presence of subsurface water.

All this evidence points to a vast saltwater ocean lying below the icy crust of Europa. An important issue is how deep we would have to penetrate through the ice to reach it. Knowledge of the thickness of the ice is critical not just for geology, but also for future exploration of the moon, which could ultimately involve placing landers with a capability to drill into the ice. The nature of the icy crust can be explored by studying past asteroid impacts on Europa. When craters form on Earth's Moon or Mars, for example, a peak usually appears in the evacuated centre, due to material being thrown up after the object hits the surface, in a kind of rebound effect. Scientists have studied the few craters on Europa that are large enough to have central peaks, one of which, called Pwyll, is 26km (16 miles) across and has a peak in its middle about half a kilometre (a third of a mile) high. The impact of the asteroid that formed Pwyll did not completely melt through

Europa viewed from a distance of 677,000km (417,900 miles) by the *Galileo* spacecraft. The ice-covered world is almost the size of Earth's Moon, with a diameter of about 3160km (1950 miles). The long, dark lines reveal fractures in the surface crust that are thousands of kilometres long and as much as 40km (25 miles) wide.

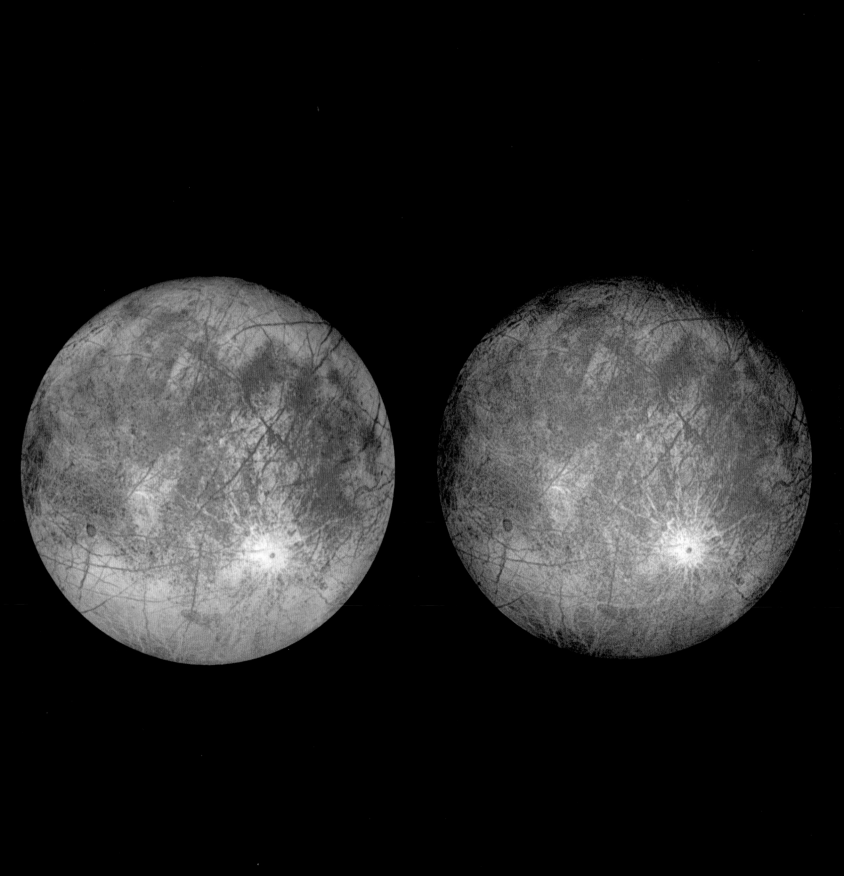

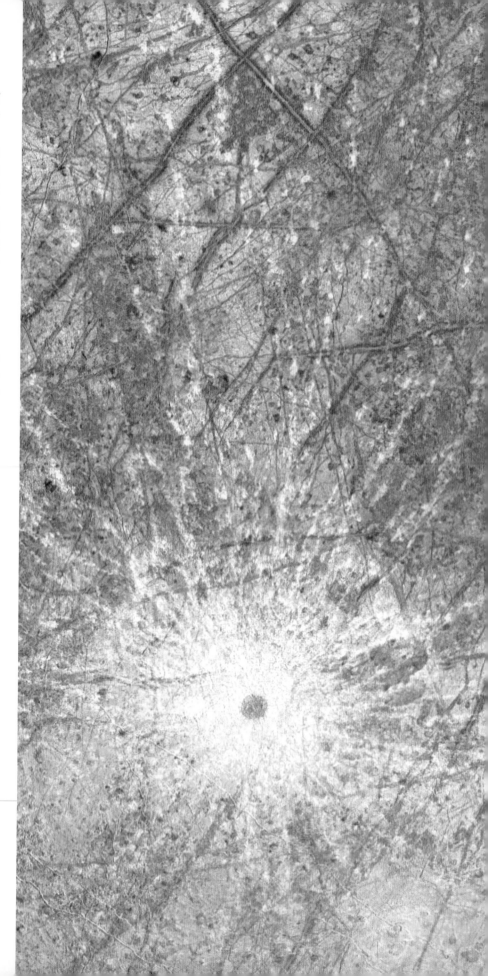

the upper ice layers. Using computers to simulate the energy released by such impacts, geologists have calculated that the surface ice on Europa must be at least 3–4km (2 to 2.5 miles) thick.

No one really knows the full depth of the ocean below this ice – it may extend as much as 100km (60 miles) down, with a slushy upper layer and very salty liquid water below. The floor of the ocean marks the start of the rocky interior, most likely eventually leading to a relatively small metallic core. This would make Europa's ocean much deeper than any of those of Earth; indeed, it may well be that Europa hosts the largest volume of liquid water in the entire solar system.

Warmed by gravity

Europa is four times further away from the heat of the Sun than is the Earth, making it an incredibly cold place. The temperature on its surface is -160°C (-260°F), which is why the entire body of this moon is covered in a layer of thick ice. Given these frigid conditions, how could there be liquid water below its surface? The answer is that the insides of Europa are warmed by the effects of gravity. The forces that create this underground heat are similar to those that move oceans on Earth to create tides. The immense gravity of Jupiter, coupled with that of Europa and the other Jovian moons, creates gravitational tugs, which induce tidal stresses and make Europa flex. The moon is essentially stretched and compressed by all this gravitational tugging, and the resulting internal heat helps to maintain water in liquid form below the surface ice.

This is essentially the same process that also acts on Jupiter's innermost moon called Io. Only in Io's case the tidal forces due to gravity generate the heat to drive dozens of erupting volcanoes, with vast lava flows. Indeed, Io is the currently most volcanic body known in the solar system, with remarkable plumes that can fountain to 300km (190 miles) above the surface.

These detailed images of Europa uniquely betray its sub-surface ocean. Pwyll crater (left) is 26km (16 miles) across, and is surrounded by ejecta that includes fresh water ice particles. The dark reddish regions (upper right) are thought to be the result of a complete melt-through of Europa's ice layers. The blue regions in the colour enhanced image (lower right) are composed of almost pure water ice.

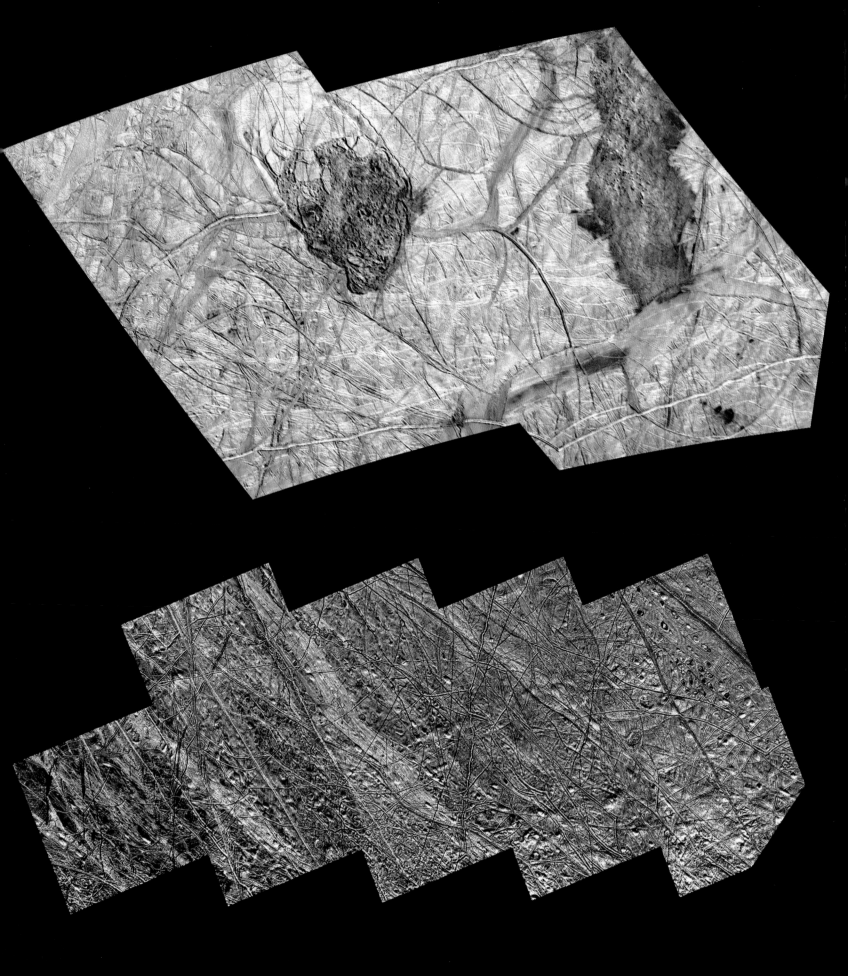

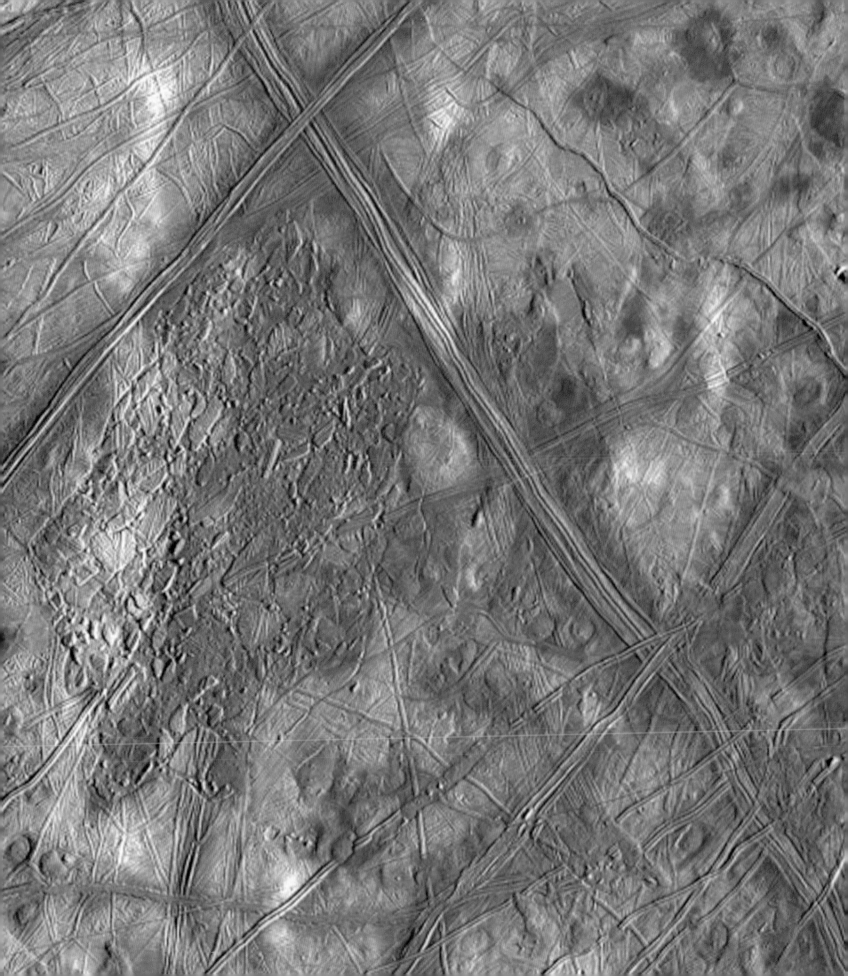

A location for life?

The possibility that Europa is harbouring vast volumes of water makes the moon an exciting subject for exobiology, or the search for life beyond Earth. As we noted earlier, liquid water is one of the essential environmental requirements for life. An additional requirement, however, is sufficient energy to support the origin of life. As sunlight would not penetrate into a subsurface ocean, life there would need to extract energy from chemical reactions and additional internal heating. One possible source of such heat could be hydrothermal vents on the ocean floor. There is evidence on Earth that primitive life forms arose near deep-sea vents, which may have been powered by volcanoes on the sea floor. The tidal heating of Europa caused by Jupiter's gravity could give rise to similar vents on the ocean floor, resulting in hot water containing vital chemical elements. Microorganisms could extract the energy from this water, setting up the path toward complex organic molecules and, possibly, life. As the vents are limited, localized energy sources, it is perhaps likely that any (past or present) ocean life would take the form of primitive single-celled organisms, found only in close proximity to these prime sites.

This close-up image of Jupiter's moon Europa was taken by the *Galileo* spacecraft. Among the intriguing features revealed are ice floes resembling those seen in the Earth's polar seas. The icy crust of Europa has fractured into plates about 30km (18 miles) across. Some of the broken plates have rotated into new positions, suggesting that they are floating on icy slush and liquid water.

One can draw interesting comparisons with subsurface ocean conditions on Earth. Lake Vostok in Antarctica is an expanse of liquid water beneath ice, covering an area of 10,000 square km (3,800 square miles), or roughly the size of Lake Ontario. It is buried under 4km (2.5 miles) of ice, which shields the lake from the freezing surface temperatures, and remarkably it seems that microbes might be able to survive in its dark and nutritionally deprived depths. Samples of ice cores taken from the lake have revealed evidence of microbiotic life, in the form of bacteria, fungi, and spore. Additional samples reveal features that are not recognizable as anything seen before on Earth.

So what are the next steps in the exploration of Europa and its oceans? There are plans to launch a spacecraft within the next decade which would orbit Europa and carry out a detailed laser and radar study of its surface. Such altimeter data could provide vital information about the oceans that lurk beneath the ice: it is expected, for instance, that a deep ocean would create regular undulations of about 10m (11 yards) in Europa's surface which would be detectable from the orbiting spacecraft. These types of studies are an essential prerequisite to placing a lander on Europa, and, ultimately, drilling or melting through the thick, icy crust to directly sample the oceans below.

Jupiter's other moons

There are other Jovian moons that could host substantial amounts of water ice on the surface, and also have mechanisms for internal heat, therefore raising the possibility of subsurface oceans similar to those of Europa. One example is Ganymede, which is the largest moon in our solar system. The crust of

Ganymede is mostly hard ice, scarred by heavily cratered regions billions of years old. The youngest craters, however, are surrounded by bright rays of freshly exposed ice excavated by the impacting comet or asteroid. More interestingly, from the viewpoint of liquid water, are the relatively younger regions of ice plains, mountains, and valleys. Some of the features seen here may also have been created by the eruption of water, caused by internal heat melting subsurface ice, which has risen like water "lava" and frozen again at the surface. *Galileo* spacecraft images have revealed long, parallel ridges, stretching for several hundred kilometres, which have smooth valley floors that may once have been flooded by such water "lava". In addition, measurements of Ganymede's magnetic field have produced similar results to those of Europa, suggesting that it too may have electrically conductive liquid material under its icy surface, in the form of a salty ocean of water, ice, and slush. In the case of Ganymede, however, the ice cover is much thicker, perhaps as much as 150km (90 miles), which would make exploration of an ocean below considerably more difficult.

Another intriguing case is Callisto, the outermost of Jupiter's largest ("Galilean") moons, whose 4840km (3010 mile) diameter is also identical to that of the planet Mercury. The mass of Callisto, however, is less than one third that of Mercury, indicating that water ice may be as abundant on the moon as rocky materials. It is thought that an ocean may lie hidden beneath Callisto's heavily cratered surface. Radioactivity at the core of Callisto is predicted to provide sufficient heat to keep the salty water from freezing. Unlike Europa, whose oceans continually fracture the icy crust, Callisto has a surface terrain that appears to have remained the same for billions of years, except for scarring due to impacts. This implies that Callisto's crust may be almost 150km (93 miles) thick, thus making it an efficient blanket to keep the interior heat from escaping.

This *Cassini* spacecraft image is one of the most detailed, global images ever taken of Jupiter. Aside from the reddish-brown bands and white ovals of the dynamic atmosphere, also visible is Jupiter's Great Red Spot, which is an extraordinarily long-lived storm.

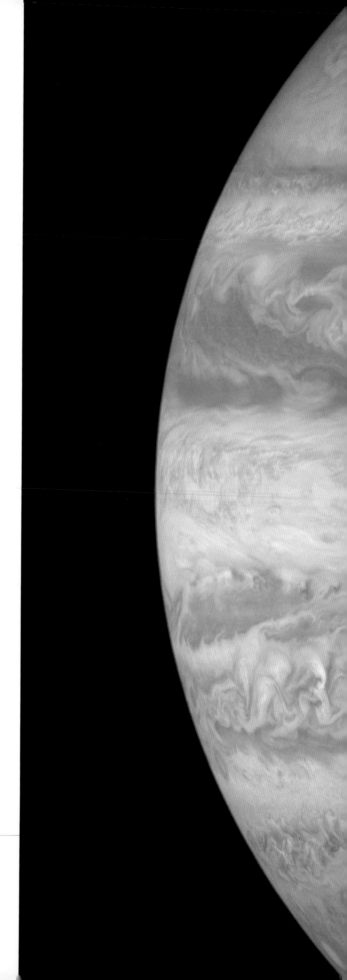

Ice on Earth's Moon

We have come to think of Earth's own Moon as a barren, dusty, and dry world. The *Apollo* astronauts of the 1960s and 1970s returned with samples of lunar rock and dirt which were completely devoid of water. Detailed laboratory analyses of these samples have also ruled out underground supplies waiting to be tapped. Unlike Mars, none of the surface features seen on the Moon is thought to have been made by the flow of liquid water in its ancient history. It was astounding then that two orbiting spacecraft in the late 1990s revealed that there is in fact water ice on the Moon after all, at the permanently shadowed, very cold regions of the north and south poles.

The first tentative evidence of this water was to come in late 1996 from the *Clementine* spacecraft. It was launched in January 1996, on a mission lasting several months, to map the Moon at different wavebands of the electromagnetic spectrum, from ultraviolet to infrared, and one of the major achievements of the mission was to chart the size and depth of a vast impact crater near the Moon's south pole. Named Atkin Basin, it is more than 2500km (1500 miles) across and almost 13km (8 miles) deep, making it the largest crater in the solar system. By coincidence the south pole of the Moon resides inside the rim of the Atkin Basin. As the Moon spins on an axis that is almost perpendicular to the plane of its orbit around the Sun, this means that the Sun always appears close to the horizon when viewed from either of the poles. For this reason large regions of the Atkin Basin remain in permanent darkness; shielded from the light and heat of the

Sun's rays for billions of years, the temperature in these areas has remained as low as -220°C (-365°F). Any ancient water ice in these polar craters may be expected to survive today.

The composition of the surface of a planet can to an extent be inferred from the way radio waves are reflected back from it. When a radio transmitter on the *Clementine* spacecraft was used to beam radio waves into the dark sections of the south pole of the Moon, the echoes were picked up using large dish antennae positioned on Earth. The key discovery was that permanently dark regions around the pole reflected radio waves in a manner consistent with the presence of ice. The same echo properties were not seen for beams reflected off other (non-polar) regions of the Moon. The conclusion was that craters at the lunar south pole contained water ice equivalent to the volume of a large lake.

This discovery of water ice on the Moon has been subsequently bolstered by NASA's *Lunar Prospector* mission, which was launched in January 1998. The aim of this low-budget mission was to prove the presence of lunar ice with greater certainty. The spacecraft went into lunar orbit with an on-board instrument called a "neutron spectrometer", which, through the detection of hydrogen atoms, sensed water ice down to a depth of about half a metre (1.5 feet) below the surface of both the north and south

This image of Earth's Moon was captured by cameras of the *Galileo* spacecraft, shortly after starting its long flight to Jupiter. The view is of dark patches of solidified lava and ancient impact craters. The Moon's north pole is located just inside the shadow zone in the top of the image.

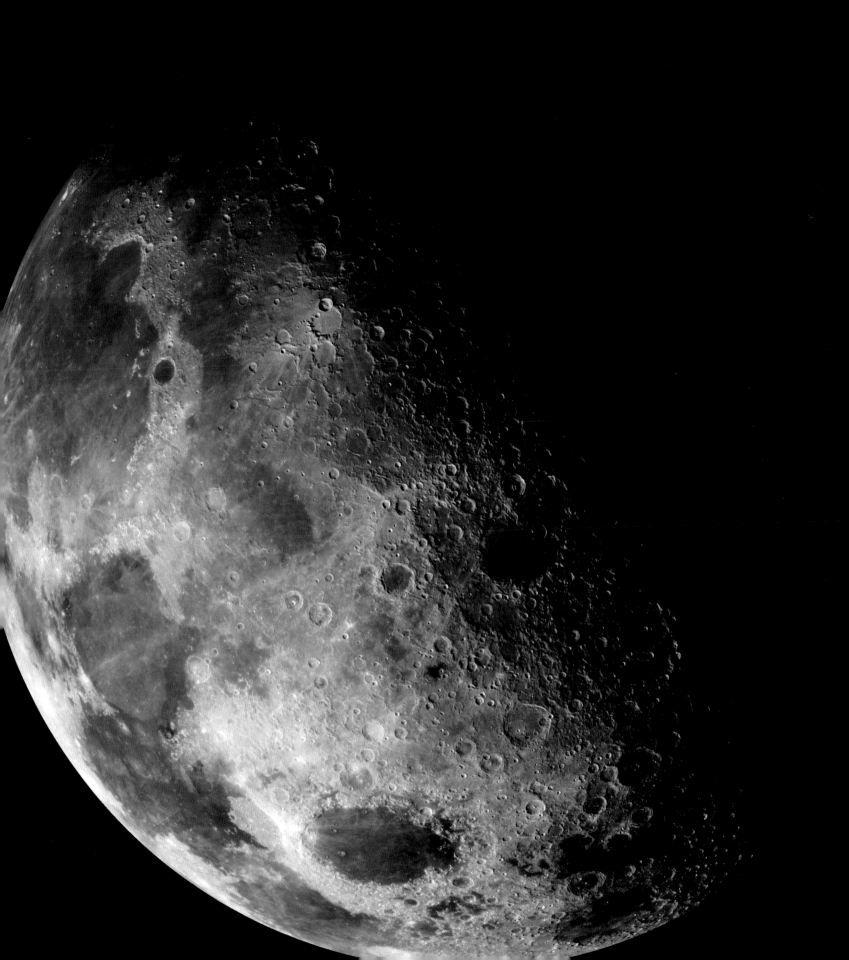

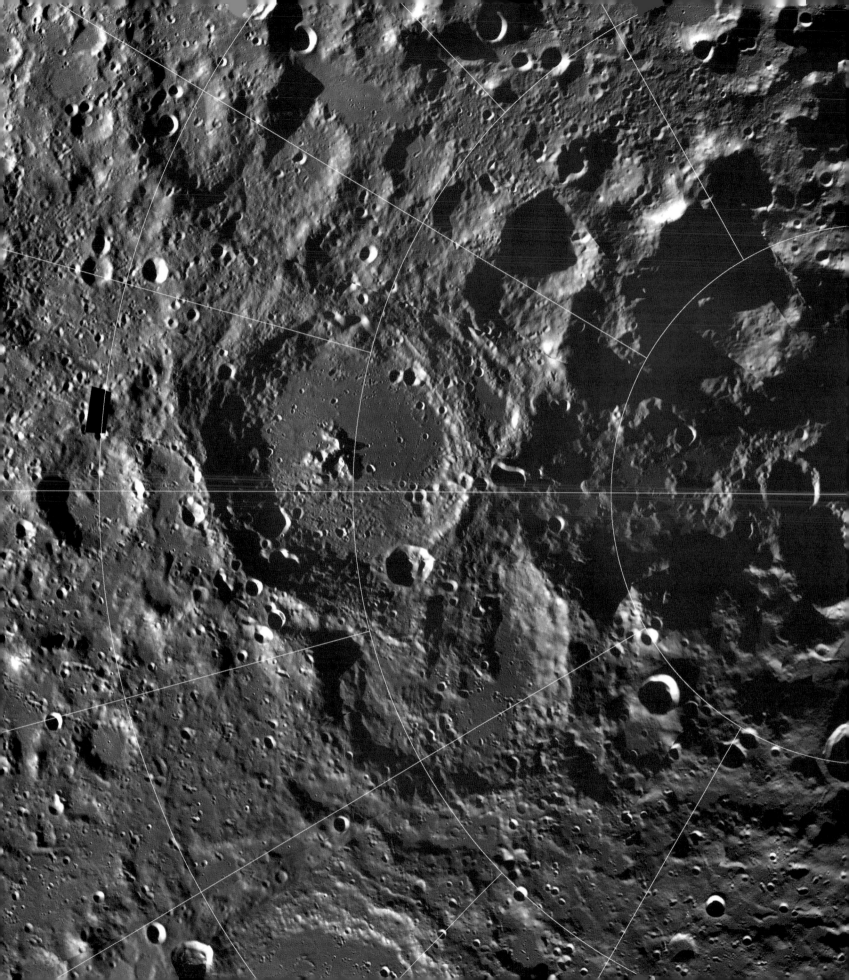

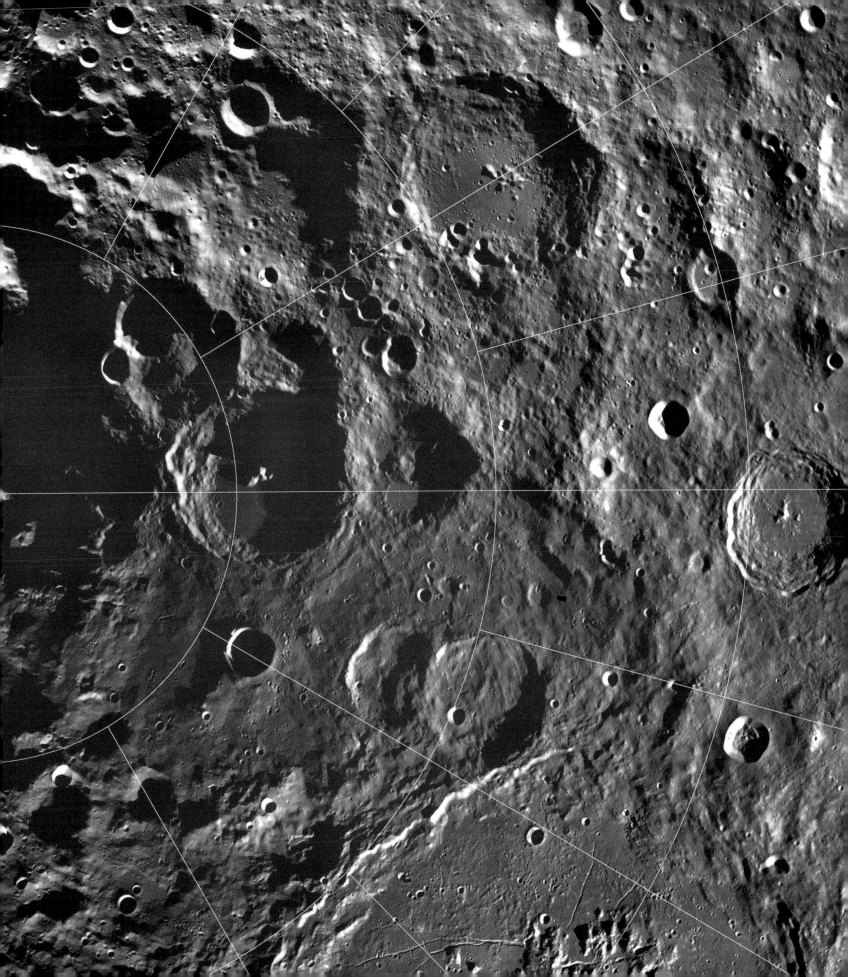

Previous pages: The south polar region of Earth's Moon is uniquely revealed in this dramatic image taken by the *Clementine* spacecraft. The extensive shadows around the pole reveal a major impact basin. A large portion of the area around the Moon's south pole is in permanent shadow and cold enough to trap water in the form of ice.

poles. Definitive signatures for the ice were confined to dark craters and overall indicated that more ice may be present at the Moon's north pole than at the south. The sum of the localized areas may be about 4000 square km (1550 square miles), amounting to perhaps a billion tons of ice. It is not found in thick layers at the surface, but is instead well mixed into the lunar soil, or regolith, in widely scattered flecks.

The *Lunar Prospector* mission ended on 31 July 1999, when it was deliberately targeted to impact into one of the permanently shadowed craters near the south pole that was thought to harbour ice. The hope was that the energy of the impact would release water vapour from the ice deposits in the soil; despite the use of powerful telescopes and other instruments on Earth, however, no plume of water vapour was detected.

The European Space Agency has now launched its first SMART-1 spacecraft to the Moon. It uses a revolutionary ion engine as the main propulsion system and is due to enter lunar orbit in January 2005. SMART-1 is carrying scientific instruments to investigate unanswered questions about the origin and nature of the Moon. Besides producing the first global lunar map in X-rays, infrared detectors will also be used to carry out new searches for signs of water ice in permanently shadowed craters.

The origin of lunar ice

The permanently dark and cold craters of the Moon have the capacity to house ice that is billions of years old, the majority of

This composite of two detailed images of Comet Wild 2 was captured by NASA's *Stardust* spacecraft during a close flyby on 2 January 2004. Jets of gas and dust surround the active surface of the comet's nucleus. *Stardust* is scheduled to return samples of the comet's dust to Earth in January 2006. The core of a comet such as this contains kilometre-sized lumps of dirty water ice, which has in the past been delivered to planets and moons by impacts during the early history of the solar system.

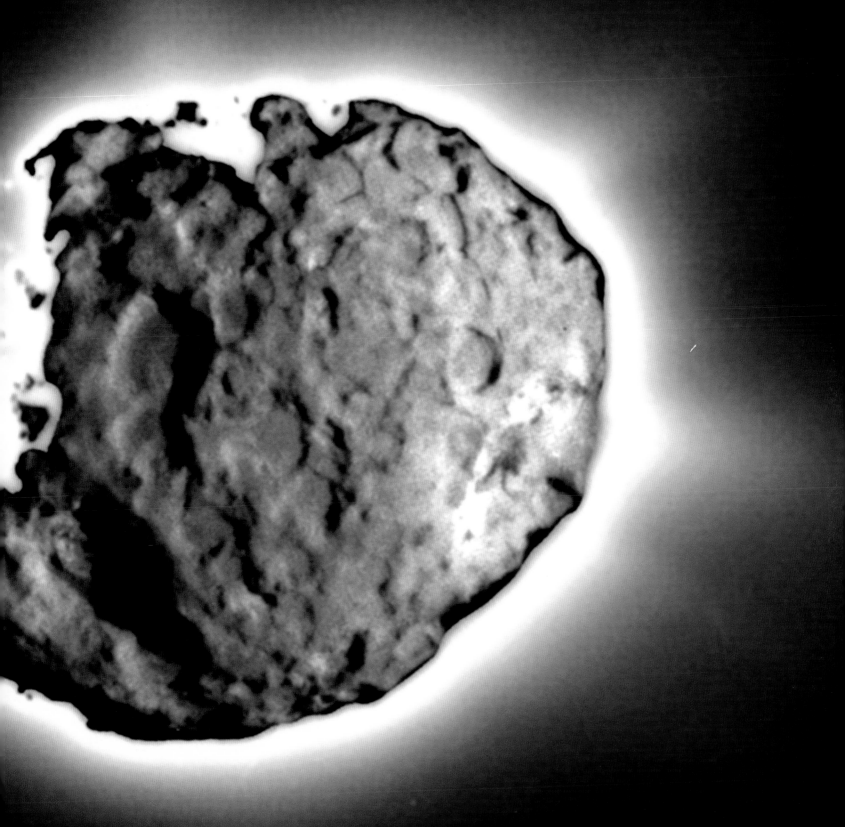

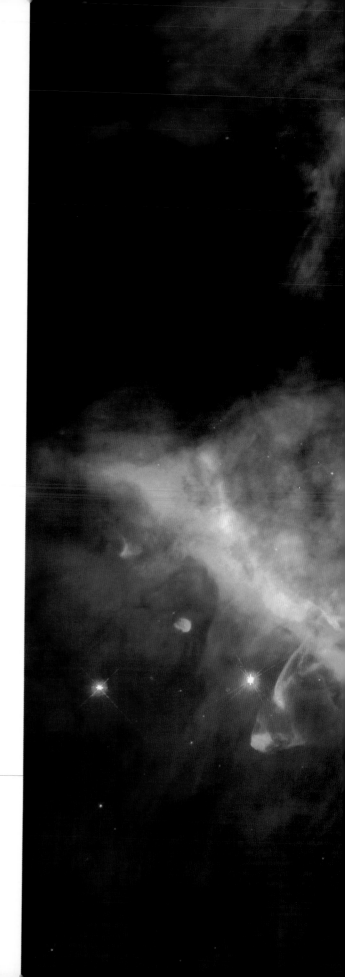

which would probably have been delivered a long time ago by impacting comets. Comets swinging in toward the Sun from the outermost regions of the solar system are mostly made of ice. They have been colliding with planets and moons for billions of years, particularly in the early history of the solar system. While most of the comets' ice would have been blasted away at the moment of impact, some of it would have remained in the lunar environment and either directly entered the craters or eventually migrated to the north and south polar regions as randomly moving molecules, to freeze. As the Moon does not have an atmosphere, its surface is also subject to bombardment by smaller objects, such as meteorites and micro-meteorites, many of which would also be coated in water ice, thereby increasing the quantity of lunar ice present.

Future exploration of the Moon

The polar sites of pristine water ice on the Moon are clearly the most promising locations for future exploration for the next generation of landing vehicles and other spacecraft. Missions that could return samples of ice back to Earth will permit the most detailed analysis and pave the way for more ambitious human exploration of our nearest neighbour in space. The discovery of substantial quantities of water has great implications for the construction of permanent lunar bases. Aside from drinking water, lunar ice can also serve as a source of oxygen which is of course vital to human life, and hydrogen and oxygen in the ice can also be used to produce rocket fuel, thereby making transport to and from the Moon considerably more economical. Permanently shadowed craters in the lunar north and south poles could eventually become the most prized real estate in the solar system beyond Earth.

Remarkably, recent studies have confirmed that water molecules also exist in the Orion Nebula (imaged here), which is a vast reservoir of gas and dust 1500 light years from Earth. Not only does the nebula provide the raw material for making new stars, but also a large concentration of water vapour has been discovered within this cloud of interstellar gas. The water contained in the comets, moons, and planets of our solar system may have been created in a cloud like the Orion Nebula.

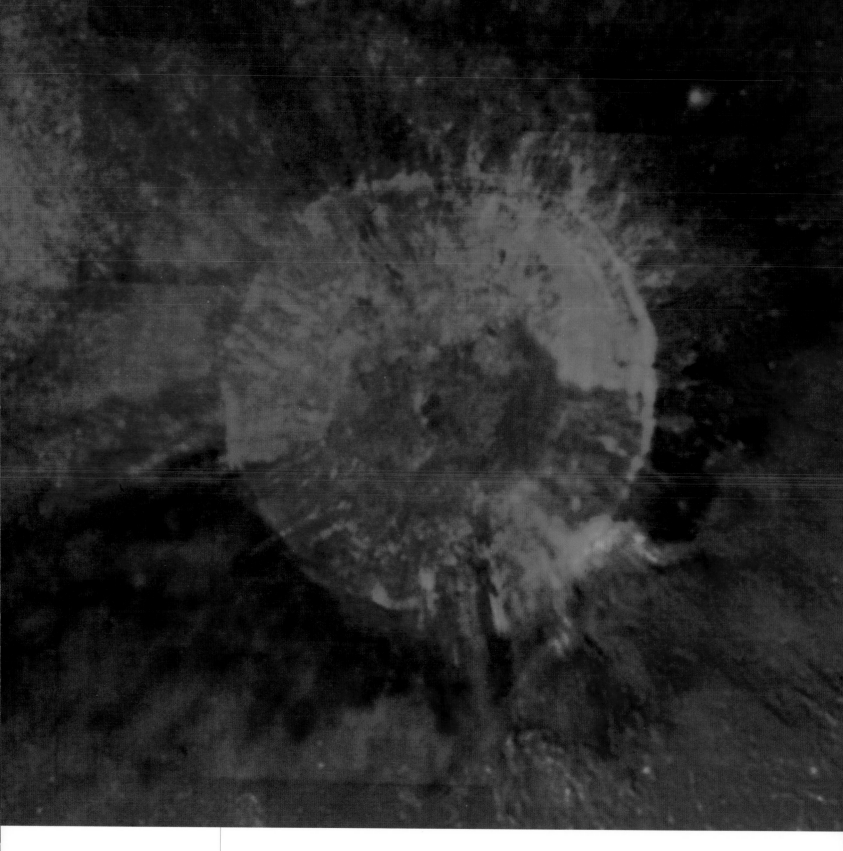

Spectacular views of the lunar impact craters Eratosthenes (right, from *Apollo* 12) and Aristarchus (left, in false-colour from *Clementine*). The very different images both reveal the detailed anatomy of an impact crater, including the floor, rim, central peak, and surrounding ejecta. Though not located in the polar regions, the shape and structure of craters like these provide insights into the plausibility of ice-containing craters on the Moon's north and south poles.

Worlds beyond our solar system

Probably since the beginning of consciousness humans have pondered the possibility of planets orbiting stars other than our Sun. Now we no longer speculate as to whether there are extra-solar planets, but instead want to know more about their characteristics and significance. For the first time in history, humans are exploring worlds outside the solar system. Since the detection of the first exo-planet in 1995, orbiting around a Sun-like star called 51 Pegasi, new planets have been discovered at the rate of more than one a month. The current tally of planets orbiting other stars exceeds a hundred, with the pace of discovery increasing all the time, and this flurry of activity has an impact not just on astronomy, but also on fields as diverse as biology and philosophy. Recent advances in the study of extra-solar planets are reviewed below, against a background where the ultimate quest is to one day discover distant Earth-like planets.

The exo-planets discovered so far

Finding planets orbiting other stars, or exo-planets, tens of light years away is a very difficult technological problem. Unlike stars,

A swirling disc of gas and dust is seen around the newly formed star called AB Aurigae. By obscuring the bright light from the star (blocked in this image by the black cross), it is possible to pick out details in the disc such as bright knots of gas and dust, which highlight the ongoing formation of planets. The disc viewed here is several times the size of our solar system.

planets do not generate their own light, but reflect the parent star's light. This reflected light, however, is very faint. In our solar system, for example, the Sun typically outshines the planets by about one billion times in visible light, so that looking back on our solar system from a distance of 30 light years, Venus, the brightest terrestrial planet, would appear almost 600 million times fainter than the Sun. Current technology does not allow the detection of such weakly reflected starlight.

As these exo-planets appear so dim, astronomers have devised other indirect techniques to locate them. The most successful method studies slight "wobbles" in the host star which are caused by the gravitational tug of planets orbiting it. This wobble is very small and demands precise velocity measurements. For example, viewed from about 10,000km (6200 miles) away the Sun's wobble, due to its family of orbiting planets, would be the distance equivalent of only a small coin. This method of study is known as the "Doppler" technique, and it uses very precise instruments called "spectrographs", which are mounted onto powerful telescopes. Spectrographs enable astronomers to measure the motion of a star as it sways back and forth, toward and away from us. Most of the known extra-solar planets have been discovered in this manner, though the results hold many mysteries. Spectrographs have also been employed to study the light from the planet-hosting stars similar to the Sun, in terms of age, temperature, and power output, mostly within a distance of fifty light years from Earth.

All the exo-planets detected so far are akin to the giant gas planets Jupiter and Saturn. Their masses range from less than a half to more than ten times that of Jupiter, and their orbital

periods around the parent star vary from around three days to 15 years. However, the planets share two surprising properties. First, the majority of them are found in orbits which are very near their star – in several cases the orbits are much closer than Mercury is to the Sun in our solar system. Star-hugging "hot-Jupiters" such as these are unexpected in the context of traditional views of how our solar system formed. The second surprise is that the exo-planets that do orbit their sun within a distance comparable to that of Jupiter are mostly in highly eccentric or oval-shaped orbits, a little like the paths of comets. Overall the planetary systems found to date are thus very unlike our own.

One of the key driving objectives of astronomy is to one day locate terrestrial planets that are similar to our Earth. Initial steps toward this major goal are finding multiple planet systems around single stars, and detecting smaller gas planets. Recent successes include the discovery of a giant planet orbiting the Sun-like star 79 Ceti. From a distance of 117 light years toward the constellation of Cetus, the tiny wobble of 79 Ceti was detected by astronomers, betraying the gravitational pull of a planet that is only 70 per cent of the mass of Saturn. The planet is in an oval orbit which brings it closer to 79 Ceti than Mercury is to the Sun. Planets in such egg-shaped orbits are likely to experience severe extremes of temperature as they move alternately closer to and further from their star. The gas planet around 79 Ceti can reach a scorching temperature of 830°C (1530°F). A multiple planet system has also been discovered around a three-billion-year-old star called Upsilon Andromedae, which lies almost 45 million light years away. Three giant gas planets orbit this star, one of which is ten times more massive than Jupiter, with an orbit that takes 1267 days to complete. A second Jupiter-mass planet orbits Upsilon Andromedae at the same distance that the Earth orbits the Sun, while a third planet is 70 per cent the mass of Jupiter, yet orbits nearly seven times closer than Mercury around the Sun, completing an orbit in just four days. Another triple planet system is known to be circling around the star 55 Cancri. Out of these three giant worlds, one is hugging the star with an orbit of just 15 days, while another is slightly further out with a

This artist's illustration shows a Jupiter-like gas planet orbiting the dwarf star Gliese 876 (depicted in the distance). The planet was indirectly detected using measurements of the star's wobble caused by the gravitational tug of the planet. Gliese 876 is 15 light years away in the constellation of Aquarius.

Previous pages: The view from a moon orbiting the Jupiter-like planet discovered around the star HD70642 is depicted in this imaginary scene. While the extra-solar planet is twice as massive as Jupiter and unsuitable for life, the lessons from the exploration of our solar system raise the intriguing possibility that moons around extra-solar planets may also harbour oceans of water beneath their icy surfaces.

44-day orbit. Interestingly, the third member of this system is a giant gas planet in a nearly circular orbit at about the distance that Jupiter orbits the Sun. The 55 Cancri planetary system could also conceivably host Earth-sized worlds, waiting to be discovered by more advanced techniques and instruments of the future.

Planet formation revisited

The majority of the known extra-solar planetary systems differ greatly from our own, and their existence is at odds with theories of how our system formed. These strange new worlds are forcing us to question our planetary origin, and even ask whether the Earth is part of a very rare family of planets. The conventional theory of the origin of our solar system holds that the nine planets (plus asteroids, comets, and other debris) formed in a flat spinning disc of gas and dust. The material was left over after the formation of the Sun at the centre of the disc. The giant gas planets arose in the outer regions, away from the Sun, where it is cold enough for dust grains to be coated in mantles of ice, which act to bind the dust particles together and slowly form planet cores, which are several times the mass of the Earth. Eventually planetary nuclei become massive enough for their gravity to capture substantial amounts of mostly hydrogen and helium gas from the surrounding disc, and make giant gas (Jupiter-like) planets. Close to the Sun there is a dearth of material and only small rocky (terrestrial) planets can form in the much warmer environment.

This conventional model for our solar system therefore predicts planets in circular orbits around the Sun, with giant gas planets in the outer solar system only. It is baffling then that the majority of extra-solar planets discovered are either Jupiter-like gas planets that are hugging their stars or are further out but in eccentric oval-shaped orbits. Standard planetary formation ideas have had to be revisited as a result of these discoveries. One theory is that newly formed massive planets may carve a gap in the swirling disc of gas and dust, dividing it into inner and outer sections. The inner disc then loses energy, causing the giant planets in the outer region to spiral in toward the star. This would explain star-hugging massive exo-planets. A consequence of the inward migration of giant planets is that their motion could lead to much smaller rocky planets being totally ejected from the solar system. In this scenario it would seem fortuitous that the giant gas planets in our solar system – Jupiter, Saturn, Uranus, and Neptune – have remained in the outer solar system: had they spiralled inward the Earth could have been ejected into interstellar space.

Over the next decade or so there will be a vast increase in the sample of known extra-solar planets. A greatly increased number is necessary to truly understand differences between planetary systems that may be at various ages in their formation history. Only then will we know whether the characteristics of our own solar system are common or exceptional.

Possibilities for life?

An exciting and profound prospect for the not too distant future is that we may be able to directly image and measure Earth-like planets travelling around Sun-like stars. Future missions such as Kepler, the Space Interferometer Mission, Darwin, and Terrestrial

Planet Finder are designed to take us toward this goal. Progress may come initially from the detection of weak light from an Earth-like planet. Using advanced spectrographs, such light can be analysed and split into its component frequencies, telling us about the chemicals and gases in the exo-planet's atmosphere.

Spectrographs could also tell us if there is life on these planets by detecting its breath – widespread microbial life on a planet will breathe and change the chemistry of the atmosphere. Ozone and oxygen, for example, have only been present in the Earth's atmosphere since blue-green algae created photosynthesis in the oceans several billion years ago. Another tracer of primitive life forms on an exo-planet may be molecules of methane. In the earliest history of Earth, microorganisms ruled the planet, gorging on energy contained in gases, and in some cases released methane. While we are a long way from the unambiguous discovery of water on a terrestrial exo-planet, its discovery light years away from Earth would be a dramatic step toward the recognition of alien life "as we know it". It is intriguing that today scientists are using the microwave analogue of lasers, called "masers" ("microwave amplification by stimulated emission of radiation"), to search for water around stars. In certain conditions molecules of water in space can emit maser beams which may be detected at a radio frequency of 22 gigahertz on Earth. These signature beams have already been detected coming from the stars Upsilon Andromedae, Epsilon Eridani, and Lalande 21185, all of which are also known to have planetary systems, according to tell-tale wobbles picked up by the Doppler technique.

It is entirely possible that life on extra-solar planets has a biology that is dramatically different from our own here on Earth.

Extraterrestrial biology based on silicon, as opposed to carbon as on Earth, would be totally unfamiliar. Life forms also might not rely on starlight, but thrive instead on geothermal energy, much like that produced around deep-sea vents on Earth, and those suspected to exist on Europa. Astronomy and astrophysics today are extending to measure chemical and biological processes around distant stars and planets. It is almost beyond our imagination what the next few decades of extra-solar planetary exploration may reveal.

The search for microbial life in the solar system and the discovery of numerous extra-solar planets has increased interest in a relatively young branch of science called "astrobiology" (or "exobiology"). As the name implies, the aim of this science is to combine the study of the universe with the study of life. The goals are not only to search for life elsewhere in the universe, but also to learn more about the origin, evolution, and future of life on Earth. New research institutes of astrobiology are being set up across the United States, Europe, and Australia, to bring together scientists from a broad range of disciplines, including astronomers, biologists, chemists, and geologists. They are motivated by several independent strands of evidence that combine to suggest biological processes may be common in the universe.

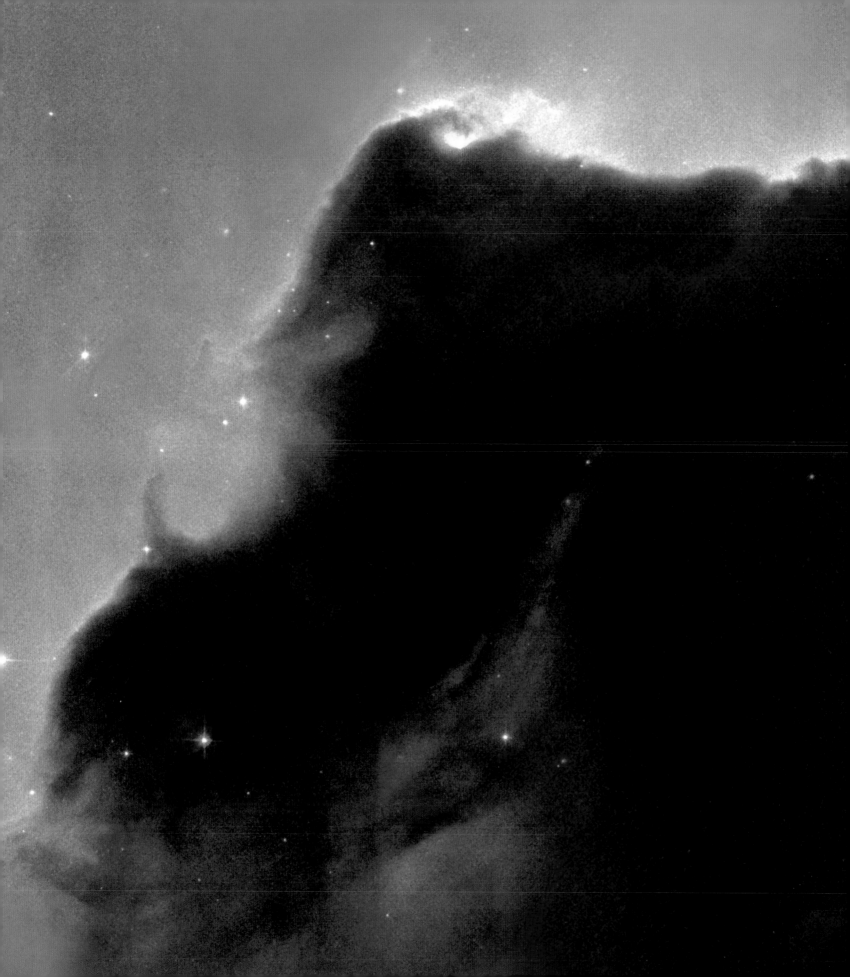

The birthplace of stars: The iconic Horsehead Nebula is captured in remarkable detail by the cameras of the Hubble Space Telescope. This cloud of cold gas and dust in interstellar space is a nursery of newly formed stars.

Origins

Powerful telescopes launched into orbit around the Earth or perched on high mountains are providing us with the most detailed and remarkable views of the universe. Backed by advanced instruments and sophisticated techniques and developments, the latest facilities are permitting exciting new advances into some of the most challenging problems in the fields of astronomy and cosmology.

This chapter explores three key areas of modern study which are linked by the common theme of origins: the birth of stars, the formation of galaxies, and conditions in the early history of the universe. The Hubble Space Telescope and other observatories are now revealing exquisite details of the processes by which stars are formed out of giant clouds of gas, with new implications for our understanding of extra-solar planet systems and the workings of galaxies. We can now also see through great distances in space to witness the evolution of an infant universe where galaxies are being assembled. This understanding of the rise of galaxies has contributed to theories about the origins of the first generation of stars and the growth of the largest structures in the universe, such as the massive clusters of galaxies.

The greatest challenge of all, however, is to comprehend the origin, early evolution, and fate of the universe itself. New microwave maps of light originally emitted just 380,000 years after the Big Bang are defining the most fundamental properties of the universe more precisely than ever before.

Several stages in the life cycles of stars are captured in this remarkable image of the giant nebula known as NGC3603. The vast clouds of star-forming hydrogen gas on the right of the image are glowing due to the intense radiation from young massive stars that reside near the centre. A cluster of more evolved massive blue stars marks the upper left of this intriguing view.

The birth of stars

An understanding of the origin of stars is vital to our understanding of the universe, as stars play a pivotal role in the structure of galaxies and the formation of planets, and are the source of life-giving chemical elements. Stars are born inside vast clouds of cold gas and dust, where dense lumps of material are gradually compressed under their own gravity. Although the basic mechanisms of star birth have been known for decades, it is only in the past few years that we have begun to witness and comprehend the finer details of this process, through the use of telescopes capable of operating in the infrared, radio, and optical (visual) wave bands .

The space between stars is not empty. Though interstellar space is more devoid of material than the best vacuum we can create in a laboratory on Earth, there is a scattering of dust and gas between the stars called the "interstellar medium". On average the gas and dust are spread very thinly, such that, if we imagine a scale model where an atom is the size of an adult, each atom (or person) would be separated by 778 million km (483 million miles), or about the distance between the Sun and Jupiter. Almost 99 per cent of the interstellar medium is made up of gas, the vast majority of which is hydrogen, then helium; the remaining 1 per cent is mostly in the form of extremely small particles of dust, each about a millionth of a metre in size. These dust particles are irregular-shaped bits of solid carbon, silicon, ice, and iron compounds, which play a vital role in the formation of planets such as Earth.

A prolific double cluster of young stars is on show here, located about 150,000 light years away in a neighbouring galaxy to ours called the Large Magellanic Cloud. The small cluster toward the bottom right contains Sun-like stars that are still forming. This image is a fine example of the interaction between gas, dust, and stars in space. Massive stars in the main cluster exploded as supernovae millions of years ago, forming the spectacular filamentary pattern seen on the left. Shock waves from these powerful blasts are believed to act as the trigger for processes that lead to the subsequent birth of new stars out of gas and dust.

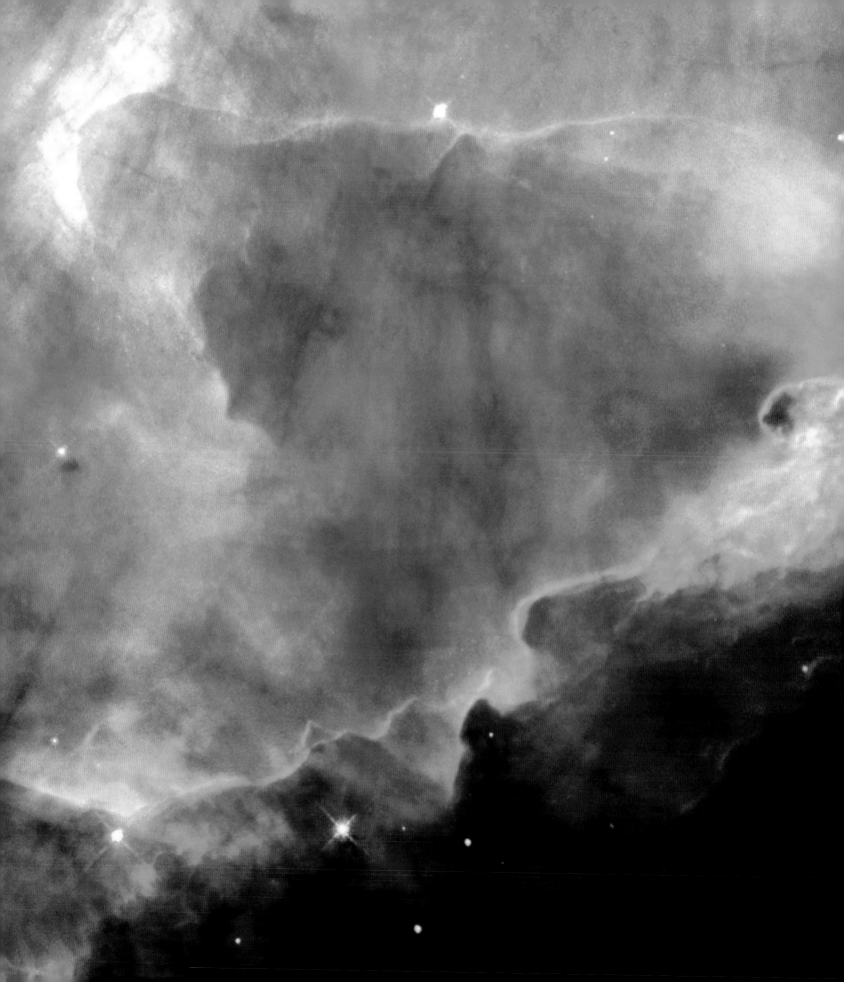

Giant molecular clouds

Material in the interstellar medium is not evenly spread, however, and can be very patchy, such that gas and dust are collected into giant clouds. These clouds, sometimes called "nebulae", consist mainly of molecules of hydrogen and helium, and can be more than a thousand times denser than the surrounding interstellar space. The nebulae in our own Milky Way Galaxy are typically a few light years across in size. The densest and most massive fragments of these nebulae are giant molecular clouds, which can contain enough gas to form more than 100,000 stars like our Sun. They are the main building blocks of our Galaxy; well-known examples of these sites of star birth can be found in the constellations of Orion, Taurus, Ophiuchus, and Chamaeleon.

Most of the gas in these clouds is very cold, with temperatures as low as −200°C (−400°F), and therefore they do not emit any optical light. This inability to see into the clouds is compounded further by the presence of dust particles. To discover more about the star-forming process, therefore, astronomers have to rely on wave bands such as infrared and radio, rather than visible light. Telescopes tuned to these wavelengths enable astronomers to see into the densest parts of dark molecular clouds, where stars are forming out of condensing clumps, knots

of hydrogen gas, and microscopic dust particles. The most opaque examples of these clumps are known as "Bok globules", which appear as sharply outlined black or dark patches against a background of stars. Embryonic stars are embedded in these light-year-sized globules, but dust particles prevent visible starlight from shining through.

When regions within these swirling molecular clouds become massive and subsequently dense, the force of gravity causes them to collapse, and continued compression raises their central temperature and pressure. After a collapse phase lasting between ten-thousand and a million years, the core temperature reaches 15 million °C (60 million °F), sufficient to ignite nuclear fusion reactions and thus power the newborn star. The onset of these nuclear reactions marks the final key phase of star formation, and the energy generated prevents any further contraction of the star.

The Hubble Space Telescope has captured some stunning images of newborn stars lighting up the shells of gas and dust that still surround them in the stellar nurseries. These views reveal nascent stars as they bathe the surrounding clouds of material in ultraviolet light. Often giant pillars of gas and dust are seen in these turbulent star-making clouds. Examples include the Cone nebula, which lies 2500 light years away toward the constellation of Monocerous, and the light-year-long pillars of the Eagle nebula in the constellation of Serpens. In both these cases, ultraviolet radiation from the hot stars erodes the edges of the dark clouds to reveal previously hidden stages of star birth. It also serves to make the surrounding hydrogen gas glow, giving rise to the red hue that colours many such stellar nebulae. The detailed

This Hubble image of the Swan Nebula shows a turbulent star-making cloud about 5800 light years from Earth. High-energy ultraviolet light from embryonic massive stars is heating the surrounding cold hydrogen gas, which glows orange-red. The region shown here is 3 light years across.

Previous pages: This star-forming cloud of gas and dust lies 2500 light years away. The head of this giant pillar is highlighted by the red halo of light due to glowing hydrogen gas. Pillars of gas several light years tall (such as this one) are incubators for newborn stars in our Galaxy.

images also reveal evidence that evaporation caused by the ultraviolet light restricts the growth of embryonic stars, by dispersing the supply of gas that might otherwise have been accumulated. This process thereby limits the amount of material available for the formation of planets in discs around stars.

Outflows and jets

One of the most startling discoveries from detailed observations of star-forming clouds is that almost every new born star seems to eject material in the form of highly heated gas. The gas either streams out in high speed jets or is blown away in an energetic wind called an "outflow". Recent observations have provided new insights into the tremendous power of this phenomenon. The jets travel at speeds of more than 1.5 million km per hour (more than 1 million miles per hour), stretching over billions or trillions of kilometres, and can carry away over this distance the mass equivalent to a hundred of our Suns. It appears that these jets are formed when gas falls onto a newborn star, becomes heated, and is blasted away, most likely aided by the magnetic field of the star. High-resolution images show that the jets originate from very close to the star. The complex knotted structure sometimes seen within these jets indicates that the gas and natal disc surrounding the young star are of an irregular, lumpy constitution.

In addition to jets, which are narrow and well confined, broader, oval-shaped winds are also seen speeding away from young stars. This is thought to be produced by the lifting of material that accumulates in the disc around the star. The outflows can extend for several light years and are powerful enough to disrupt and shape the surrounding giant molecular cloud. Using the

The beauty and mystery of space is captured here by ESO's 8-metre class telescope in Chile. A dark dusty cloud, known as Barnard 68, is seen projected against a glittering, jewel-like background of stars. The size of the dark cloud is 12,500 times the distance between the Sun and the Earth. Our knowledge of how stars form is derived from physical conditions in the cold interiors of interstellar clouds such as Barnard 68.

Hubble Space Telescope images secured over several years, it is even possible to witness how these giant flows evolve.

Jets and outflows record the recent history of infant stars clearing the reservoirs of unused gas and dust that surround them at birth. Ultimately this history can teach us a great deal about the conditions in which planets form around stars.

Proto-planetary discs

While a collapsing giant nebula provides the basis for star formation, astronomers have long suspected that stars would be surrounded by leftover gas and dust, swirling in a pancake-shaped disc. These discs are of great interest because they represent the earliest stages not only of the formation of stars, but also of planets. Sometimes known as "proto-planetary discs", they are thought to be composed of 99 per cent gas and 1 per cent dust particles, although even this tiny amount of dust makes the discs opaque and difficult to view and image. Remarkable recent advances in telescopes and instrument techniques, however, have permitted outstanding new glimpses of proto-planetary discs.

Instruments using the infrared wave band have penetrated the dusty and dark discs at various orientations from face-on to edge-on, providing information about their thickness and density. Some proto-planetary discs contain more than ten times the mass needed to make the solar system. They can be divided into two main categories: thick discs associated with proto-stars prior to the onset of nuclear fusion in their cores, and thin debris discs around older stars with much hotter cores. The formation of planets takes place during the transition between these two types of discs. The Earth and other planets in our solar system are thought to have formed out of discs such as these, about 4.5 billion years ago. Indeed, proto-planetary discs seem to be commonplace and, if planets are a natural by-product of star formation, the expectation must be that planets are equally commonplace.

We are on the verge of learning a great deal more about these planet foundries, thanks to new facilities such as NASA's Space Infrared Telescope Facility (SIRTF).

This computer-generated simulation shows a young star forming out of a surrounding disc of gas. The central star gradually accumulates material from the disc until its core becomes hot and dense enough to begin nuclear fusion reactions.

The dawn of galaxies

Galaxies are the building blocks of the universe. There are more than a hundred billion of them in the observable universe, each a cornucopia of millions or billions of stars, gas, and dust, all held together by their collective gravity. Although there is tremendous variation in their size, brightness, and structure, galaxies are usually divided into three broad types: spirals, ellipticals, and irregulars. Spiral galaxies, such as our own Milky Way, are so-called due to the magnificent spiral arms that extend outward like a pinwheel from the central nucleus. Elliptical galaxies have an oval shape and, unlike spirals, have long since ceased making new stars. Irregular galaxies have a chaotic appearance with no discernible shape or structure, and are usually sites of prolific star formation. A holy grail of modern astronomy is discovering how all these different types of galaxy were assembled in the early history of the universe.

The quest to discover how galaxies formed is one of the most open in astronomy and cosmology. We saw earlier in this chapter that astronomers can observe the formation of a star because it is a current and ongoing process – our own Galaxy has been making a few Sun-like stars every year for the past ten billion years. We have, however, never directly seen nearby galaxies forming, essentially because they were all made billions of years ago in a gradual, not instantaneous, process. Astronomers have to use advanced telescopes to observe the faintest and most distant galaxies, thus probing back in time to their infancy. Although light travels at an incredible 300,000 km per second (186,000 miles per second), the universe is so vast that it can take billions of years for light to reach us from these most distant galaxies. Thus, the further away a galaxy is, the further back in the past we see. The deepest views of the universe are allowing us to see faint primeval galaxies as they were more than ten billion years ago. As a result, we are beginning to witness the dawn of structure in the universe, such as galaxies and groupings of galaxies. In addition to using telescopes as "time machines" in this way, our knowledge of the birth of galaxies can also be advanced by detailed studies of the present-day populations of those galaxies which are nearby. These local galaxies contain clues as to how they have been sculpted.

Building blocks in deep space

The highly sensitive Hubble Space Telescope has provided us with some of the most detailed views ever obtained of the early universe. The most remarkable of these images show views dating back almost 13 billion years to when the universe was barely a billion years old. Among the recognizable shapes of spiral and elliptical galaxies, these deep observations reveal thousands of

A magnificent view of the grand Whirlpool Galaxy. Luminous young stars and glowing hydrogen gas trace out the elegant spiral arms that define the galaxy. Intricate dust lanes are also revealed in detail, providing astronomers with fresh insights into how spiral galaxies are assembled.

scarce infant galaxies, at an epoch when they started to form stars in significant numbers.

Looking along corridors that are more than ten billion light years long, faint smudges of blue light can be seen, which turn out to be significant. These are the proto-galaxies that formed the galaxies we see around us. Numerous distorted and dwarf pre-galaxies, which appear to be merging and growing in a "bottom-up" hierarchical scheme, consist of about a billion very young and hot blue stars. They each measure barely 2000 light years across, tiny in comparison to the 100,000-light-year span of the disc of our (ordinary) Milky Way Galaxy. While thousands of these star-forming, peculiar galaxies can be seen in the early history of the universe (up to ten billion years ago), they are virtually absent in the modern epoch. It is thought that these proto-galaxies served as the building blocks of galaxies we see today. Detailed Hubble images have provided direct evidence to support the theory that large galaxies grew from these small fragments coming together, rather than from the collapse of gigantic clouds of gas. Through repeated collisions and mergers, the fragments were transformed into the regular elliptical and spiral galaxies we see relatively close to us today. According to this scenario we might expect that our Milky Way Galaxy still contains components of this "bottom-up" assembly. They may be part of the globular clusters, which represent tight collections of the oldest stars residing in the extended halo of our Galaxy, or part of the central bulge which contains more than half the mass of our Galaxy.

The evolution of galaxies is a complex subject involving some of the most challenging issues which currently face astronomers. A fundamental quest is to understand the events in the early universe that caused hydrogen and helium to clump into the clouds destined to form galaxies.

Help from nature's telescope

The fragments that come together to assemble grand galaxies are very small and incredibly distant, which makes them

A classic spiral galaxy called NGC4603 is captured by the Hubble Space Telescope. At 108 million light years from Earth, this is the most distant galaxy in which special stars called "Cepheid variables" have been individually picked out. These pulsating stars provide a vital tool for measuring distances in the universe. (The bright star near the centre is a foreground object in our Galaxy, unrelated to NGC4603.)

This is the deepest view of the universe ever obtained in visible light. Known as the Ultra Deep Field, this Hubble Space Telescope image samples galaxies formed 13 billion years ago, when the universe was barely 5 per cent of its present age. Using a one-million-second exposure, the field shown contains a zoo of almost 10,000 galaxies. Studies of this image will improve our understanding of the rise of galaxies in the early universe.

extremely hard to see with even the most powerful space-borne or ground-based telescopes. Fortunately a trick of light called "gravitational lensing" can act to create a very powerful natural telescope that comes to the aid of astronomers.

Gravitational lensing is a remarkable feature of Einstein's theory of general relativity. The effect is that the presence of a substantial body curves or distorts space and acts to bend the path of light. In space it is possible for astronomical objects, such as galaxies or groups of galaxies, to bend light in this way, and under the correct conditions multiple images of a single distant source can be seen in the background of a much closer lensing galaxy.

An intriguing example of a gravitational lens is the cluster of galaxies called Abell 2218 (see Chapter 4), which lies almost about two billion light years away. Abell 2218 plays the role of a foreground lens, bending light and making objects aligned behind it appear more than 30 times brighter. One of the objects seen magnified in this manner is a collection of a million stars at a remarkable distance of 13 billion light years. This infant cloud of hydrogen gas and young stars is thought to be a building block in the formation of galaxies. Gravitational lensing by the Abell 2218 galaxy cluster has allowed astronomers to detect the birth of primeval fragments that later merged to build galaxies. These fragments are thought to have been present in great numbers in the early universe, more than ten billion years ago.

The farthest galaxy currently known was discovered by the combined power of a 8m class telescope in Chile and the amplification of light by a (foreground) gravitational lensing cluster of galaxies three billion light years away. The newly claimed record

Origins

73

A phenomenal burst of star formation is revealed in this image of a galaxy called NGC4214, located 13 million light years from Earth. The spectacular central feature is a cluster of hundreds of newly formed massive stars, each at least 10,000 times more luminous than the Sun. Starburst galaxies such as NGC4214 provide clues as to how populations of stars first "switch on" in galaxies.

announced in March 2004 places the infant galaxy at an astounding 13.2 billion light years from Earth. The faint light detected by this method left this galaxy when the universe was only 4 per cent of its current age.

Local analogues

Aside from capturing direct views of faint pre-galaxies in the early universe, it is also possible to learn about galaxy formation by studying the features of those that are local, or relatively nearby. By viewing these neighbouring galaxies, astronomers can compare their shape and properties to their distant counterparts, and the identification of common features permits faraway infant galaxies to be more easily recognized and understood.

Of particular interest are any relatively nearby dwarf galaxies, which are a fraction of the size of normal galaxies such as the Milky Way. These objects, mostly irregular in shape, are characterized by vast reservoirs of unused gas and high rates of star formation; in effect they have not evolved very far, making them similar to primeval galaxies. Combined with their proximity, this makes the dwarfs very useful for inferring the conditions in young galaxies in the early universe, and astronomers are looking to understand their star-making history. One such example is NGC1705, which is an irregular dwarf galaxy about 17 million light years away. Using the Hubble Space Telescope it is possible to actually pick out individual stars, thousands of which are seen lighting up its central regions. It is thought that NGC1705 experienced a very large increase in its star formation activity about 30 million years ago, which is recent compared to its estimated age of 13 billon years. The subsequent evolution of

its population of stars will determine the chemical composition and physical shape of NGC1505 .

Among the best analogues to galaxies during their formation are a special breed of galaxy called "starbursts". The defining characteristic of a starburst galaxy is that it is experiencing a period of intense star formation activity, during which it can be forming more than a hundred Sun-like stars per year, and has a correspondingly high rate of supernovae explosions. This activity, normally triggered by collisions or mergers with another galaxy, akin to the encounters expected between faint blue proto-galaxies in the young universe, makes starbursts tens to hundreds of times more prolific than normal galaxies. The enormous stellar energy released can dramatically alter their structure. Starbursts are also some of the most luminous galaxies, as the majority of their newly formed stars are powerful and massive.

Starbursts and other active local galaxies provide clues as to how the very first stars lit up the universe. It is likely that the universe made a significant fraction of its stars in a violent firestorm of star birth, barely a few hundred million years after the Big Bang. While stars continue to be formed in galaxies today, their present birth rate may be insignificant compared to this tremendous firework display which characterized the early universe.

As galaxies go, this beautiful spiral galaxy called NGC1512 is relatively nearby, at a distance of 30 million light years from Earth. The bright core of the galaxy is surrounded by a remarkable ring that hosts the vigorous formation of new stars. Studies of NGC1512 are helping astronomers to understand starbursts in much more distant, very young galaxies.

Ripples from the Big Bang

The quest to understand the formation of stars and galaxies leads us to the ultimate challenge of discovering the origin and early evolution of the universe itself. This is the subject of a science called "cosmology", which raises some of the most profound questions. Did the universe have a beginning sometime in the past? How old is it? What is its structure and shape? How will the universe continue to change in the future? By studying the oldest light in the universe in unprecedented detail, astronomers have provided exciting new insights into some of the most fundamental questions of cosmology.

An ancient glow

The current, widely accepted model for the origin of the universe is the "Big Bang" model. This cosmological picture simply states that the universe initially expanded from a very hot and dense origin, and it is still stretching today. As a result of this continued expansion, the universe is now much cooler and more tenuous. Support for the notion of Big Bang cosmology comes from Edwin Hubble's discovery in the 1920s that all galaxies are rushing away from each other, and the fact that the primordial amounts of light elements such as hydrogen and helium predicted by the theory were subsequently discovered to be correct.

In the distant past, and very close to its origin, the universe would have been extremely hot and very densely packed with subatomic particles. At a time when it was one hundred millionth of its present size, the temperature would have been more than 270 million°C (520 million°F). The radiation or light from this time constantly scattered off particles such as electrons and could not travel freely or escape, as if trapped in thick fog. However, around 380,000 years after the Big Bang origin, the universe had expanded and cooled sufficiently for electrons to be locked into hydrogen atoms. As there were no longer vast numbers of free electrons to continually scatter light, the paths of photons of light were no longer impeded. The photons were therefore able to travel more freely and directly: the "fog" had now lifted, and the universe was transparent after this epoch.

The radiation released at this time still fills the universe today and is known as the "cosmic microwave background". This afterglow of the Big Bang is the oldest light in existence, and it is highly significant because it carries an imprint of the infant universe. Originally released in the form of gamma rays, as the universe expanded it cooled to lower and lower energies, and today we detect it as microwave radiation. In 1992 the NASA satellite Cosmic Microwave Background Explorer (COBE) accurately measured the temperature of this radiation as −276°C, (−465°F, more commonly known as 2.73 Kelvin). The universe today is bathed in this frigid radiation.

An artist's impression of NASA's Wilkinson Microwave Anisotropy Probe (WMAP) leaving the Earth/Moon system and travelling 1.5 million km (1 million miles) into space. From that remote vantage point the mission mapped the microwave background radiation of the universe in unprecedented detail.

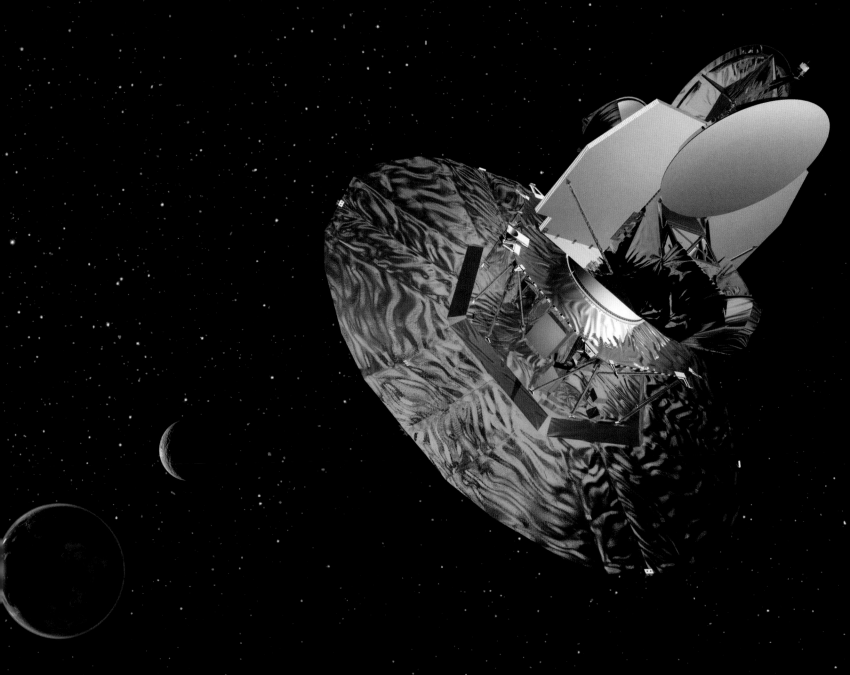

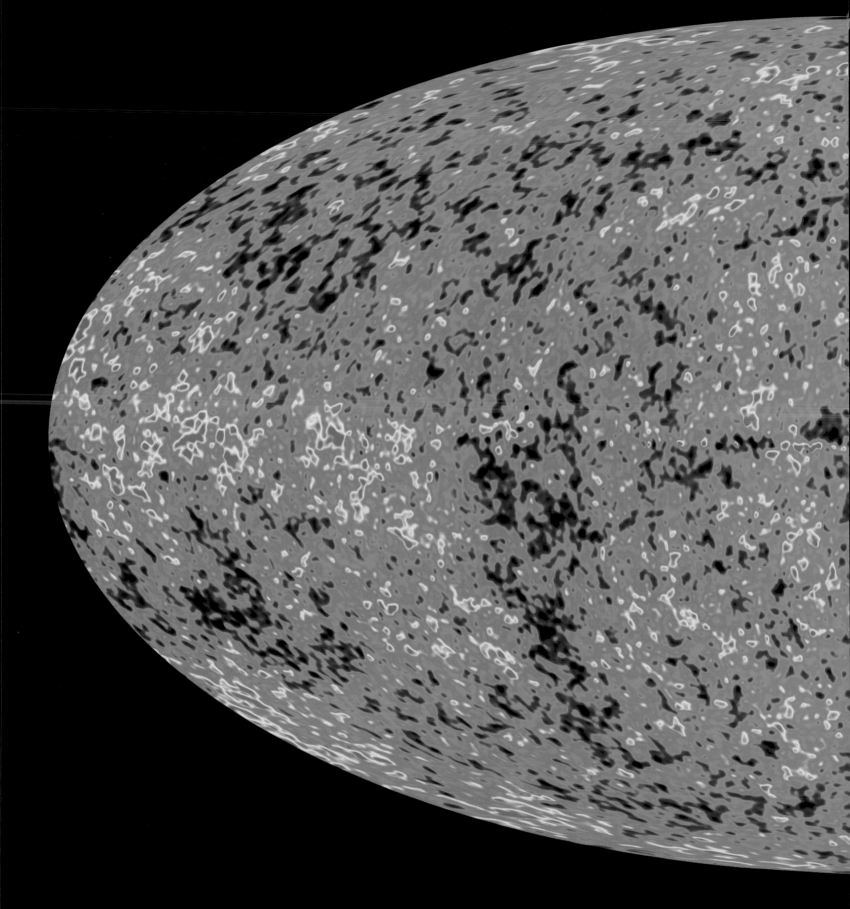

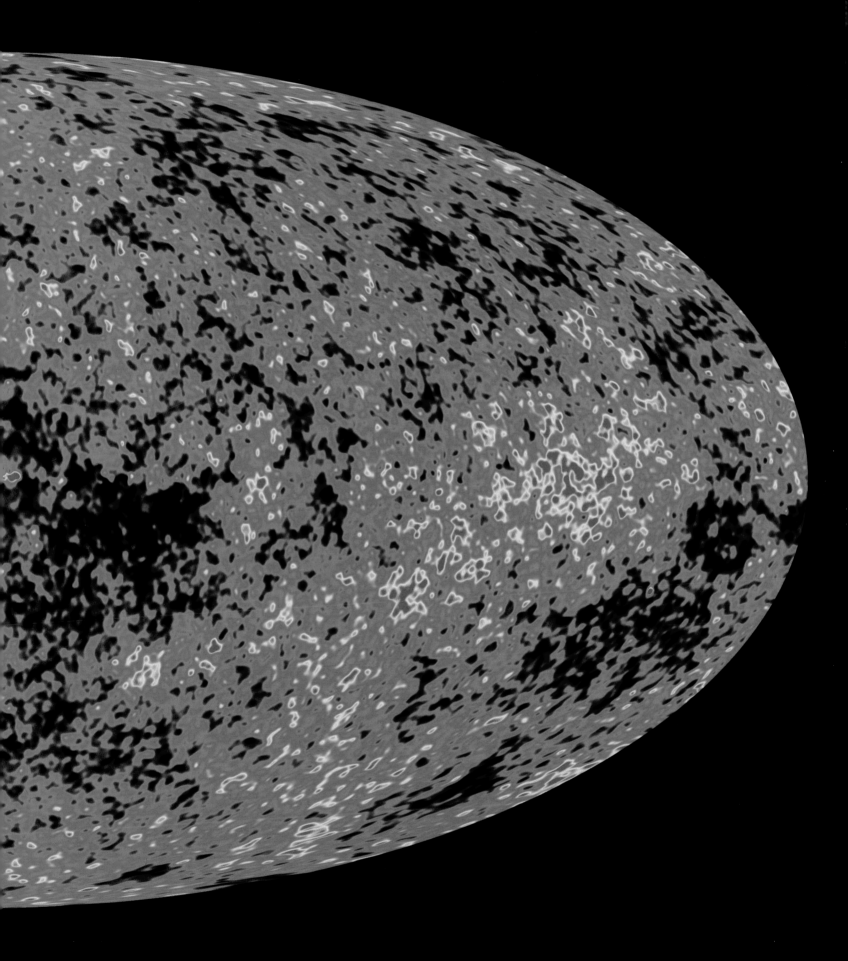

Previous pages: This microwave map from WMAP is
a high-resolution all-sky picture of ancient light that
emerged a mere 380,000 years after the Big Bang.
The colour differences reveal minuscule temperature
variations which in turn reflect conditions that must have
been set in place a tiny fraction of a second after the Big
Bang. The patterns imprinted in this map correspond to
the "seeds" that grew to become the clusters of galaxies.
The patterns also encode critical information about the
age and geometry of the universe.

Portrait of an infant universe

Cosmic microwave radiation is a vital relic of the newborn
universe – it is the equivalent of examining a picture of an 80-
year-old person taken on the day of his or her birth. It can also be
thought of as a fingerprint that contains information about the
composition and conditions of the early universe. In order to
decode this fingerprint, astronomers have captured the most
detailed, highest resolution images ever taken of the cosmic
microwave background.

Launched on 30 June 2001, NASA's Wilkinson Microwave
Anisotropy Probe (WMAP) satellite embarked on a three-year
mission to survey the entire sky and produce a sharply focused
map of the microwave radiation. Orbiting about 1.5 million km
(1 million miles) from Earth, WMAP has revealed stunning details
of extremely tiny variations in the temperature of the radiation.
The minute variations had already been detected by the COBE
mission in 1992, but WMAP provided data of much higher resolution and sensitivity, and has been able to resolve temperature
changes which vary by only one millionth of a degree. The pattern
of these tiny temperature fluctuations betrays information on the
seeds that generated the large structures of the universe from
individual galaxies to super-clusters of thousands of galaxies
stretching millions of light years.

The microwave map produced from the WMAP mission is an
oval projection of the entire sky, similar to the way the globe of
the Earth is often projected in an atlas. It contains a fossilized
record of the impact of matter and energy on space, barely
380,000 years after the Big Bang, long before stars and galaxies

existed. So what can we learn about the universe from these
maps? Theories of the amount and nature of mass and energy in
the universe make specific predictions about the size of the temperature fluctuations in the background radiation, and their
extent in the sky. The "fingerprints" revealed in the WMAP observations can be compared to predictions made by different
theories to see which one matches the best.

Although these new data will be scrutinized by scientists for
many years to come, there have already been some startling new
revelations. First, the present age of the universe is more precisely dated at 13.7 billion years old, rather than the previously
estimated 12 to 14 billion. Some of the light in the cosmic
microwave radiation pattern is polarized, which means that,
instead of vibrating in all directions like normal light, it is limited
to vibrate in a specific direction. This is a trait of starlight and its
presence in the WMAP data is a signature of the energy released
by the first stars in the universe. It seems that the first generation
of stars in the universe ignited only 200 million years after the Big
Bang, which is much earlier than had been previously thought.

The WMAP data also suggest that we are made of very rare
stuff indeed! That is, ordinary matter in the form of atoms, which
makes up only 4 per cent of the universe, including all the billions
of stars and galaxies. The remainder of the universe is in more
exotic forms, namely the enigmatic "dark matter" and the equally
mysterious "dark energy", which we will return to later in this
book. This component of matter and energy does not radiate and
cannot be observed directly using telescopes, although it does
have a gravitational influence. It appears that 23 per cent of the
universe consists of dark matter, which can take some exotic

forms such as subatomic elementary particles called "wimps" and "axions". Even more mysterious is the remaining 73 per cent of the universe that is now thought to be composed of dark energy. No one really knows what the form of this dark energy is. One possibility is that it acts like a type of "anti-gravity", whereby the effect is to make objects repel (unlike gravity) and therefore make the universe expand faster. While gravity holds planets, stars, and galaxies together, dark energy tugs on the fabric of space and time. This strange new repulsive force would push galaxies apart faster and faster. A physical basis for dark energy remains to be worked out. It does not easily fit in with current theories of physics, and we do not yet know what causes it. It may have been created at the very earliest moments of the universe, at a time when the other forces of nature such as gravity and nuclear forces came into being.

The shape of the universe

The fact that 96 per cent of the universe is in the form of dark energy and dark matter has a direct impact on its shape, and its fate. The amount of matter and energy spread across space gives the universe its overall shape, known in cosmology as "curvature". If the density of all forms of matter and energy in the universe is greater than a critical value, then space has a positive curvature and the universe is said to be "closed". If you imagine shining two powerful lasers out into space in a closed universe, the initially parallel paths of the laser will eventually (over billions of light years) converge, cross, and return to their starting point. Alternatively, if the average density of matter in the uni-

verse is less than the critical density, the curvature is negative and the universe is "open". The path of the imaginary lasers would become further and further apart as they move across the universe. Finally, if the average density of the universe is exactly equal to the critical density, then we have a flat universe with zero curvature. The light from our laser beams will now remain perfectly parallel even after travelling billions of light years. By placing very tight limits on the angular size of the slightly hotter and cooler regions in the cosmic microwave radiation, the WMAP observations have determined that the universe is indeed flat with zero curvature. The principal implication of this is that the universe will expand forever: the galaxies, it seems, will continue to race away from each other. There remain many uncertainties, however, one large caveat in this picture of the future of the universe being that the nature of dark energy is unknown. If for some reason dark energy changes in some significant way with time, then it could have a correspondingly dramatic effect on the evolution of the universe.

The WMAP mission has shown that detailed images of cosmic microwave radiation can provide remarkable insights into the nature of the universe. There is no doubt that these studies have tremendous additional potential. Beyond the WMAP data, there is the prospect of more sophisticated missions to come, such as the European Space Agency's *Planck* spacecraft, due to be launched in 2007. It will provide even tighter constraints on the properties of the cosmic microwave radiation and thus on the infant and nearby universe.

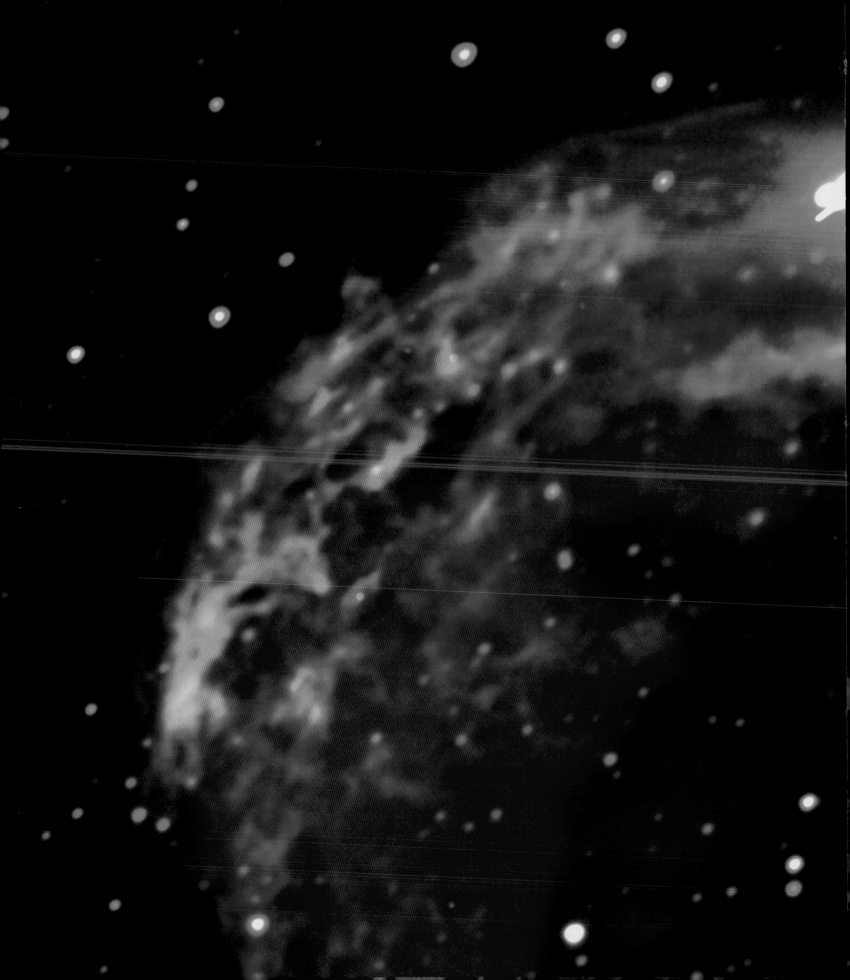

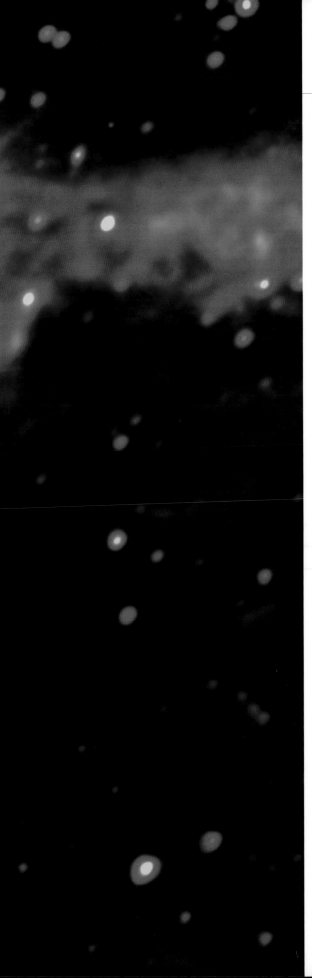

Racing toward a violent end: Powerful winds from a massive star called HD 192163 have sculpted this vast gaseous shell. After a prolific and relatively short life, the luminous star will explode as a supernova at an age barely one-thousandth that of the Sun.

Cosmic fireworks

Gazing up at a clear night sky, the universe may seem like a tranquil and serene place, but this is largely an illusion created by the vast distances and scales of space. We will see in this chapter that the universe is in fact continually being shaped by very energetic events, involving huge amounts of matter and immense forces. Beginning with our own backyard star, the Sun, scientists continually monitor powerful flares and giant solar eruptions that can strike the Earth's atmosphere with hazardous effect. Indeed, the final acts in the life stories of stars are marked by spectacular explosions and the ejection of enormous shells of hot gas, spreading life-giving stardust across space. Peering through giant telescopes we can also capture detailed snapshots of violent collisions between galaxies, creating vast streamers and triggering bursts of new star formation. A journey toward the centre of our Milky Way Galaxy takes us past fierce explosions and the emission of high-energy radiation, to a black hole of around three million times the mass of the Sun. Such super-massive black holes are the prolific central engines of enigmatic cores of active galaxies called "quasars", which can be seen in the outer reaches of the universe. Our knowledge of all these energetic phenomena is being advanced by new studies of the intense energy released in wave bands beyond the visual; the branch of X-ray astronomy, in particular, has recently widened our view, with space observatories such as NASA's Chandra satellite producing the most sensitive views ever seen of the cosmic fireworks on display.

This view of what is known as the Pencil Nebula shows the wispy remnants of a massive star destroyed in an ancient supernova explosion some 11,000 years ago. The shock wave from the blast is so powerful that gas in the Pencil Nebula is still moving today at more than 645,000 km per hour (400,000 miles per hour).

The explosive Sun

When viewed from the Earth the Sun can appear placid, calmly yielding a constant stream of heat and light. It is, however, a dynamic and stormy place, capable of powerful eruptions and high-energy outflows that can generate significant effects on Earth. Observed from a distance of 150 million km (93 million miles), the Sun is still the only star that we can study in great detail and on the finest scales. The latest instruments can detect and resolve intricate loops and bubbles of hot gas that are only a few hundred kilometres across. Such precise study of the Sun and its activity not only provides a key platform for understanding the nature of stars in general, but also enables scientists to report and ultimately predict the conditions in the local environment around Earth. These surroundings are known as "space weather".

The Sun, like any ordinary star, is a vast thermonuclear reactor. The energy to make the Sun shine is generated by the fusion of hydrogen nuclei into helium in its ferocious central core, which is approximately a fifth of the radius of the Sun in size. In the core, gas is squeezed to a pressure that is 200 billion times that on the Earth's surface, and the temperature rises to a remarkable 15 million °C (60 million °F). The energy released in this nuclear crucible is transported outward by high-energy photons of light that continually collide with gas in the layers surrounding the core. These photons do not reach the outer layer of the Sun, called the "photosphere", for hundreds of thousands of years; once the photons have arrived at the photosphere, they escape as visible (white) light. The photosphere has a temperature of 5700°C (10,300°F) and marks the principal surface layer of the Sun that we see from Earth.

A cycle of activity

Energetic activity of the Sun is sporadic, and is usually signalled by the temporary appearance of various transitional phenomena. These transient events include sunspots, explosive flares, and giant eruptions that carry away vast amounts of electrified gas called "plasma". Such phenomena are not always present, but instead occur in a "solar cycle" which is closely related to the magnetic properties of the Sun. Magnetic field lines are threaded through its upper layers, such as the photosphere, and can become very twisted over a period of several years because different parts of the Sun rotate at different rates. This spread of rotational speeds is critical. Unlike a solid terrestrial planet such as the Earth, the Sun is a gigantic ball of gas that spins more rapidly at the equator than at its poles. As the gas circles the Sun faster at the equator, it pulls the magnetic field lines into a twisted and knotted pattern; the more distorted or kinked the field lines, the more active the Sun becomes. It takes 11 years to

An image of the active Sun from the SOHO spacecraft showing broad arches of dense, electrically charged gas. Known as "prominences", the dynamic structures are suspended in the Sun's tenuous corona by strong magnetic fields.

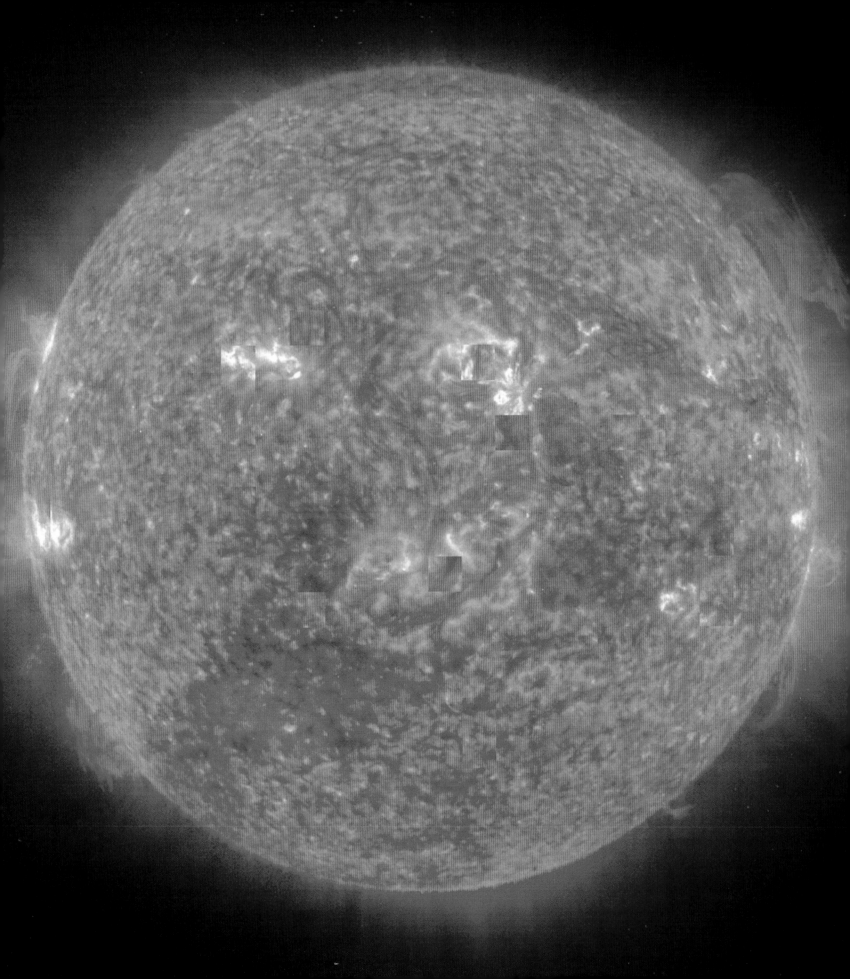

complete a full solar cycle, during which the magnetic pattern of the Sun goes from a simple untwisted one (when the Sun is not exceptionally active) to a highly contorted arrangement (when the Sun is at its most active), then back to a quiescent state. The last peak of activity was between late 2000 and 2001, and is now declining toward a minimum around 2007.

Over the past few years a suite of satellites and telescopes has provided remarkable new images of stormy events on the Sun. A fine example is SOHO, the Solar and Heliospheric Observatory, which is jointly operated by the European Space Agency (ESA) and NASA. Launched on 2 December 1995, it occupies a permanent vantage point for viewing the Sun, at a distance of approximately 1.5 million km (930,000 miles), or about four times the distance between the Earth and the Moon. SOHO is observing the Sun in unprecedented detail, providing exciting new insights into the outer solar layers and the stream of plasma that blows through the solar system.

Spots and flares

Typically the first indication of stormy activity on the Sun is the appearance of dark spots on its surface, that is the photosphere. Called "sunspots", these are Earth-sized patches that migrate around the Sun over days or weeks as the star rotates, and are dark in appearance because they are about 3000°C (5400°F) cooler than the surrounding layers of the photosphere. They are essentially regions of intense magnetic fields, which tie up most of the energy in localized areas, and lead to a lowering of its temperature. Sunspots usually occur in groups that can last for weeks or even months. Remarkable close-up views of them have recently been obtained by the Swedish Vacuum Solar Telescope, located at La Palma in the Canary Islands. These high-definition images clearly reveal the activity of the magnetic fields and the detailed features of the spots, including dark pores and intricate thread-like structures.

Large, prominent sunspot groups are often accompanied by flares, which are the most explosive phenomena linked with the

Fountains of electrified and incredibly hot gas are revealed rising from the Sun in this detailed image from the TRACE spacecraft. The loops of gas extend 480,000km (300,000 miles) into the Sun's corona and span the distance of 30 Earths at their base.

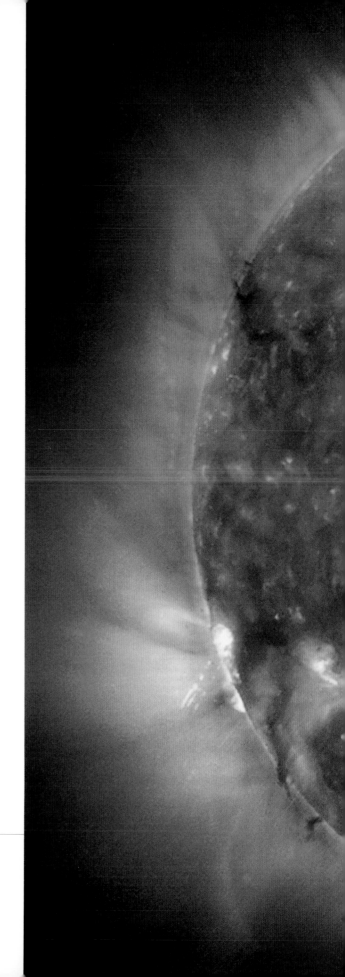

Sun. Appearing as bright eruptive patches, flares occur when the magnetic energy built up near a sunspot is heated to millions of degrees in a few minutes. The energy released in a single flare is equivalent to a billion tons of TNT, and bursts out in many forms, including X-ray and gamma-ray radiation, energetic particles such as electrons and protons, and outflows of gas. Solar flares are classed by how bright they appear in X-rays. The biggest and most powerful ones are labelled X-class; those with a tenth of the emission of X-class flares are called M-class, and the weakest types are C-class. During the Sun's active period the X-class flares can occur several times per year.

On 28 October 2003, the most powerful X-class flare recorded over the past 50 years erupted on the Sun. Although the Earth's magnetic field acts like a shield against the most harmful effects of radiation and energetic plasma, the resulting geomagnetic storm struck the Earth less than a day later, signalled by the appearance in the night sky of beautiful aurorae, or wispy curtains of coloured light. The storm caused power surges in electrical grids across the United States and Canada, and interfered with the operation of communication satellites used for cellular phones. The very high radiation levels also posed a threat to astronauts working in the orbiting International Space Station.

Bubble and trouble

The largest flares on the Sun are commonly the trigger for enormous eruptions of material that speed across space into the solar system. Known as "coronal mass ejections" (or CMEs), they are vast bubble-shaped bursts of super-heated gas. By following their expansion and rise away from the Sun, space missions such as SOHO are providing remarkable new insights into them, and into the explosive nature of the Sun. Each eruption is characterized by the release of almost 100 billion kg (220 billion lb) of extremely hot solar plasma, which speeds away from the Sun at 1.6 million km per hour (one million miles per hour). Expanding to a size greater than the Sun itself, the energy tied up in these gigantic bubbles of electrified gas can be a hazard if their path

Bright solar flares are captured in this image of a stormy Sun taken in ultraviolet light by the SOHO spacecraft. The explosive flashes convert enormous amounts of magnetic energy to launch material into space at millions of kilometers per hour.

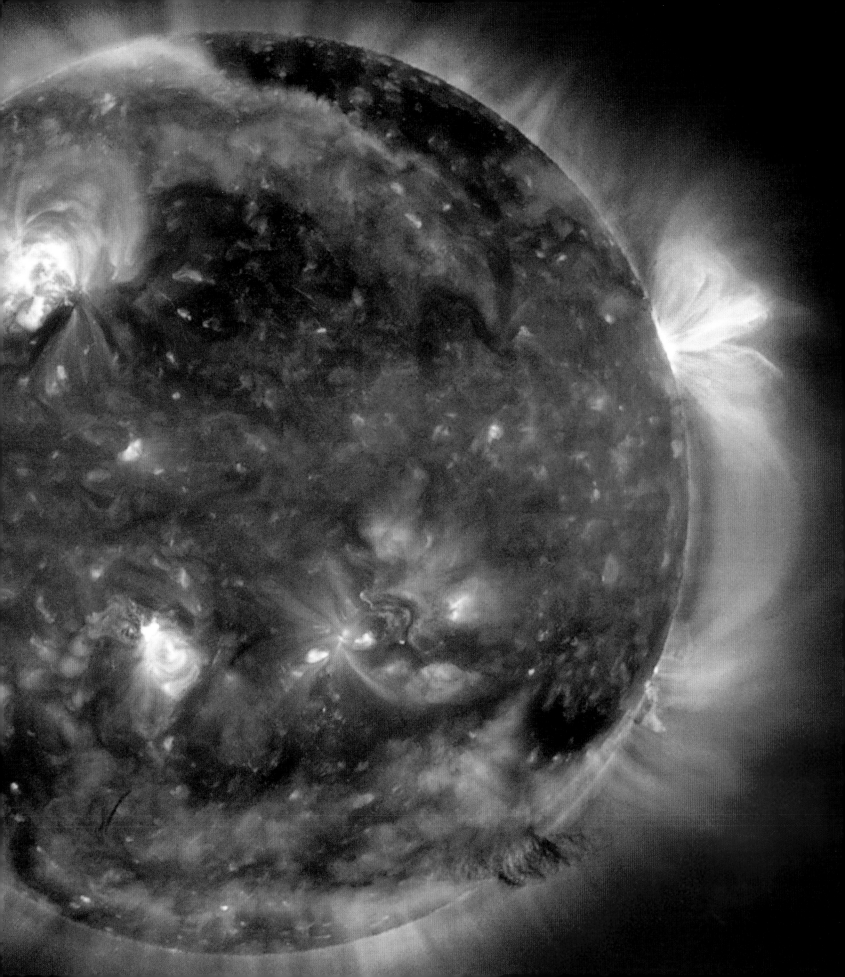

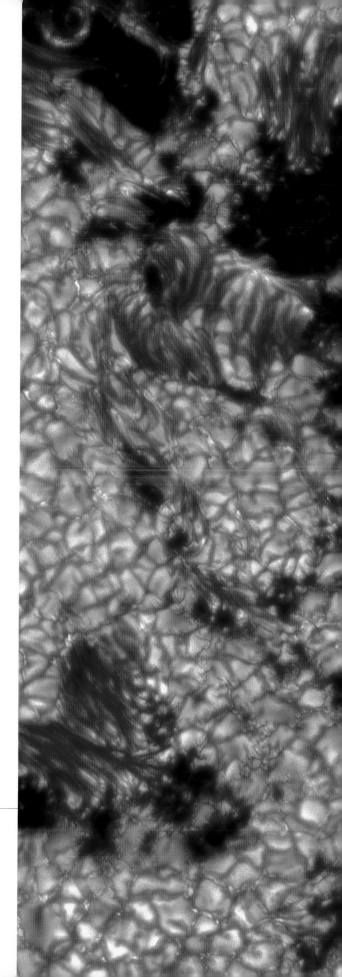

intersects the orbit of the Earth. A coronal mass ejection propelled away from the Sun on 6 January 1997 struck the Earth four days later, causing permanent damage to a new orbiting communications satellite. The intense radiation and plasma burnt out intricate electrical circuits and disabled the entire satellite, causing widespread television blackouts across North America. Coronal mass ejections are most frequent close to the peak of the 11-year solar activity cycle, and are capable of dumping huge amounts of damaging electricity into the Earth's atmosphere.

As we are hugely dependent on communications satellites and electrical power supplies, scientists now use telescopes on satellites such as SOHO to continually monitor conditions on the Sun, in order to provide advance warning that a coronal mass ejection or material from a powerful flare is heading toward us. Such notice of an impending solar magnetic storm would enable satellites to be shielded, by safely switching off sensitive components and computers. Monitoring space weather in this way is also vitally important to the safety of astronauts and high-altitude pilots.

Flows and waves on the Sun's surface

There is a growing wealth of data on the variable and dynamic conditions at the surface layers of the Sun (the photosphere), and this in turn contributes to our understanding of the structure and properties of the interior regions, which can never be directly viewed. The Sun's surface is a turbulent boiling pot of 6000°C (10,800°F) plasma that is in constant motion, due principally to the rotation of the Sun itself, which carries the equatorial region around at a speed of 2000m per second (6500 feet per second). Its mottled or granular appearance is due to pockets of hot gas that rise up from the interior. This "convective motion" happens on two principal size scales: larger cells of bubbling material each typically 1000km (600 miles) across, with a lifetime of barely five to ten minutes before they are replaced by new material from the interior layers, and "super-granules" which stretch over 20,000

Taken during July 2002 from the Swedish Solar Telescope, this is the most detailed image ever obtained of the Sun. The darkest regions in the centre are the umbra of planet-sized sunspots, which are surrounded by intricate thread-like structures. Features on an unprecedented scale of only 100km (60 miles) are revealed which betray the turbulent upper layers of the Sun.

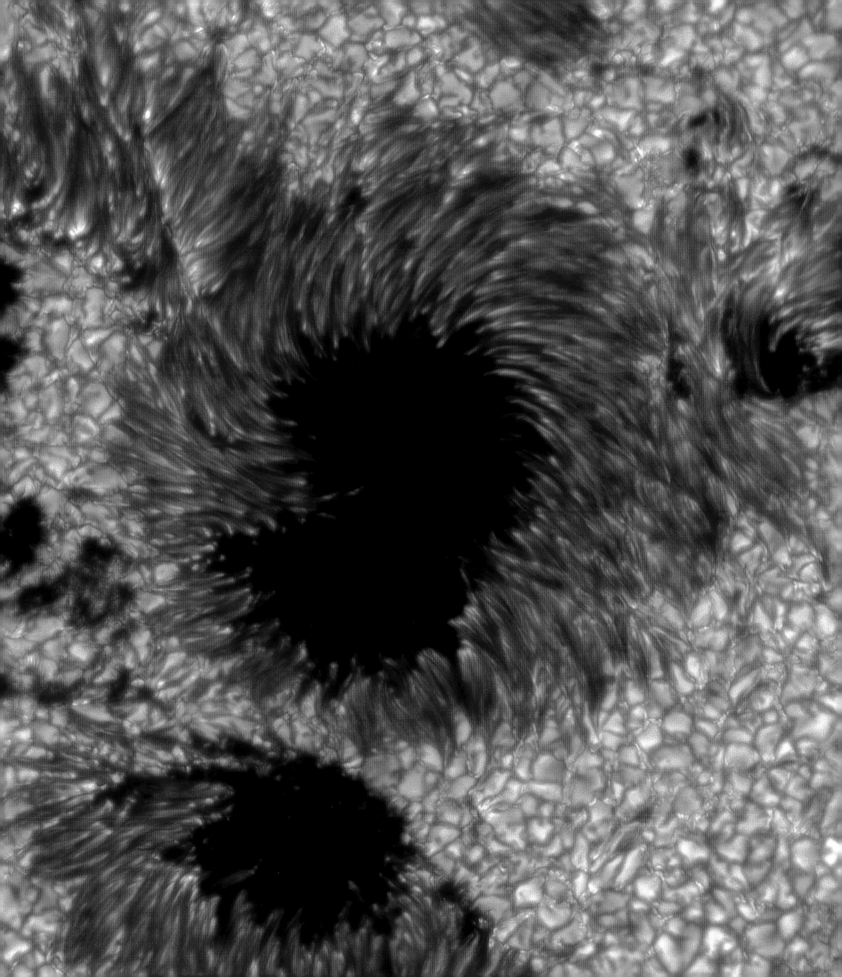

From left to right, this sequence of images spans two hours and show the eruption of an enormous coronal mass ejection (CME) on the Sun. Billions of tonnes of electrified gas are being propelled into the solar system. On this occasion the CME was not headed toward Earth.

km (12,500 miles) and last about ten to 20 hours. Together these two types of granular cells form a substantial part of the normal surface layer of the Sun that we see every day.

Another property of the Sun is that it is continually pulsating, with patches of its surface vibrating up and down over a time period of about five minutes. This oscillation is driven by sound waves that are generated by pressure deep below the surface. These waves, and the frequency of the vibrations they produce, can be used to probe the interior of the Sun, in the same way that geologists use seismic waves from earthquakes to learn about the Earth's core. Not only has this study, called "helioseismology" led to discoveries about the temperature and structure of the Sun's inner regions, but also its techniques have been extrapolated to apply to the internal workings of other stars.

Climate change

Recently, scientists have suggested that long-term changes in the Sun's brightness could have an equal, if not greater, influence on global Earth temperatures as does rising emission of carbon

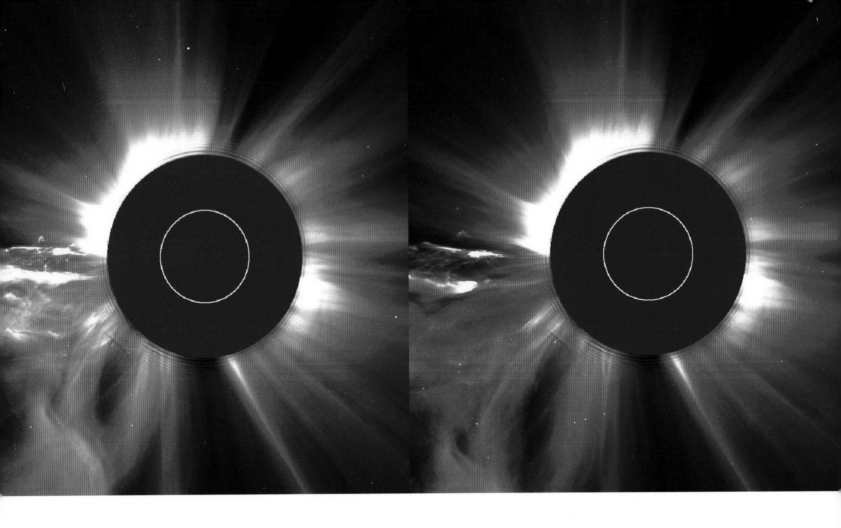

dioxide from cars and factories. An intriguing historical connection between the temperature on Earth and long period changes in the Sun's activity comes from the study of tree rings and ancient layers of glacial ice. Scientists have found a curious link to a "little ice age", so named because of a marked decrease in temperatures at the start of the 13th century. Having dropped to 2°C (35.6°F) below the long-term average, temperatures did not recover until the 18th century. This "little ice age" time period also corresponds to an exceptional era of apparently lowered solar activity, including sunspots.

Another significant factor for climate change may be the way variations in the radiation from the Sun affect the Earth's ozone layer. Ozone naturally occurs in the atmosphere and acts to protect life on Earth from much of the Sun's harmful ultraviolet radiation. The effect of prolonged periods of enhanced activity on the Sun, including flares and coronal mass ejections, would be to bombard the Earth's upper atmosphere with vast amount of electrically charged particles. The greater influx of particles would break up molecules of gases such as nitrogen, the atoms of which

can then break down the ozone molecules and, in turn, would thin the ozone layer that buffers the Earth. A vastly thinned or disappearing layer of ozone would expose Earth's planetary life to dangerous doses of ultraviolet radiation.

The energy emitted by the Sun also directly and indirectly-drives winds and clouds on Earth. Solar heat produces temperature differences on Earth that subsequently lead to pressure differences, making air move from high to low pressure regions, which creates winds. Additionally, the powerful magnetic field of the Sun protects the entire solar system from cosmic rays, which are very energetic particles that travel though space. A long-term fall in solar activity would mean a significant drop in the strength of the magnetic shield, and a greater amount of cosmic rays would be allowed to enter Earth's atmosphere. Recent satellite data have revealed a link between the amount of cosmic rays that reach the Earth and the extent of low clouds. A long-term increase in cosmic rays could reduce the cloud cover and increase the amount of solar radiation that reaches the Earth's surface, which would lead to a rise in global temperatures.

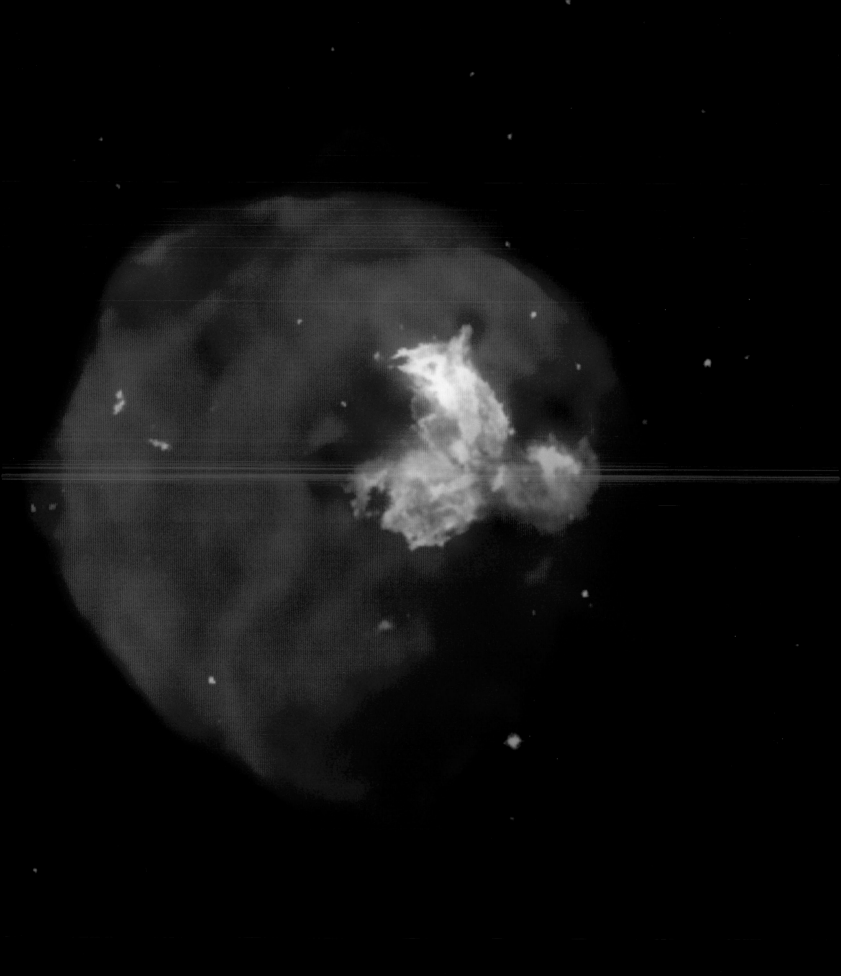

The death throes of stars

The stars and constellations in the night sky have appeared to be almost unchanging over the entire period of recorded human history, but stars are not eternal. We have already seen earlier in this book that stars are born in nurseries within giant clouds of gas and dust. Their birth is followed by an active life and an ultimate demise that concludes a story spanning billions of years, and this death of stars is one of the most fascinating topics of modern astronomy. While the birth of stars is generally an orderly process, their death can be extremely violent and catastrophic. The final stellar acts are not only spectacular sights, but also fundamentally important processes by which galaxies are enriched with the chemical elements that are critical for life itself. During its life a star will go through many changes in its physical state, some of which are only partially understood, and astronomers are therefore striving to gain fuller knowledge of this evolutionary process.

The life of a star is characterized by a continual battle between the force of gravity trying to collapse it and the counteraction of pressure from hot gas in the star's interior trying to support it, and the most stable phase in this life is the period called the "main-sequence", when these forces of gravity and pressure are balanced. Our own Sun is currently in such a phase and, fortunately for us on Earth, will continue to be so for about another five billion years.

The interior heat and radiation that support a star are provided by nuclear reactions in its core. There are many different and complicated reactions possible, but the basic one which occurs in stable Sun-like stars is the fusion of hydrogen nuclei into helium. Eventually, however, a star will exhaust the supply of hydrogen fuel in its core, along with any other elements available for fusion energy generation, which means that there will no longer be any source of heat to counteract the force of gravity. A period of drastic change will follow, and some of these remarkable phases have been beautifully captured in recent years by powerful telescopes.

Shells and bubbles

The specific evolutionary path a star follows, and in particular what happens to it when its nuclear fusion energy source is exhausted, depends critically on how massive it was at birth. An ordinary "lightweight" star such as our Sun will evolve by greatly expanding its outer layers and turning into a red giant. This will happen about five billion years from now, when it will bloat outward to engulf the planet Mercury and possibly even Venus. The

The X-ray glow from a shell created by the destruction of a massive star. The powerful shock wave generated by the supernova explosion more than 2000 years ago has heated the stellar material to more than ten million °C (18 million °F). The X-rays imaged by the Chandra satellite are combined with optical (green) and radio (red) images to provide a unique insight into the drama of this cosmic violence.

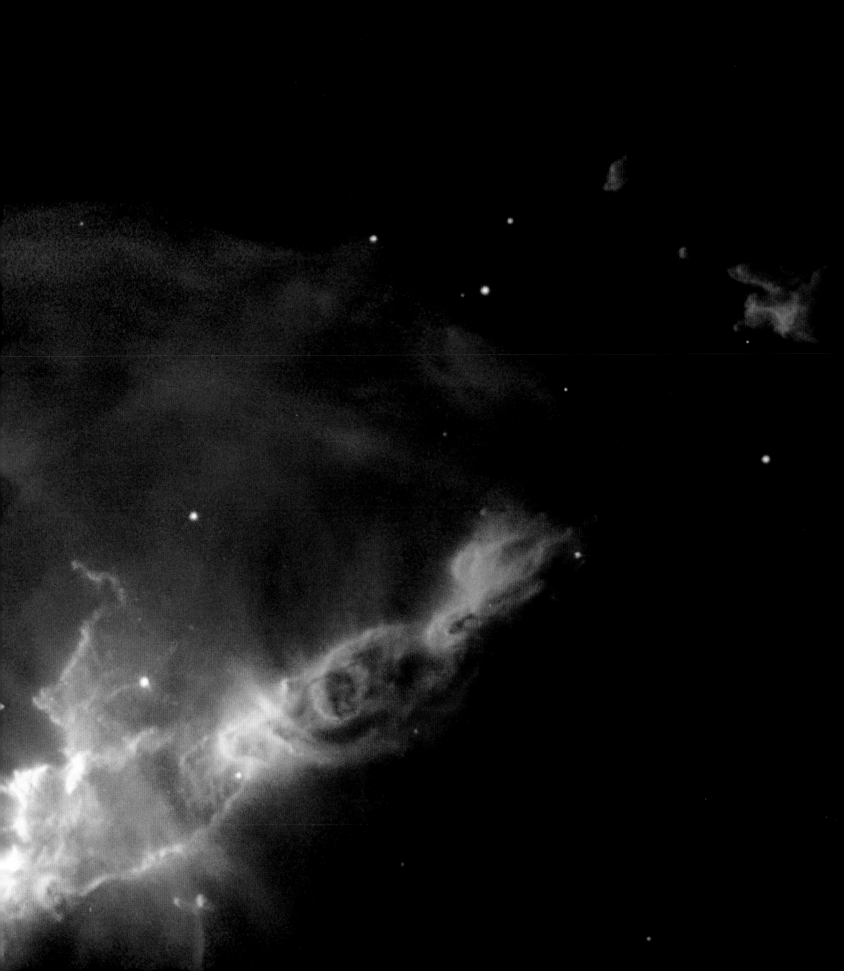

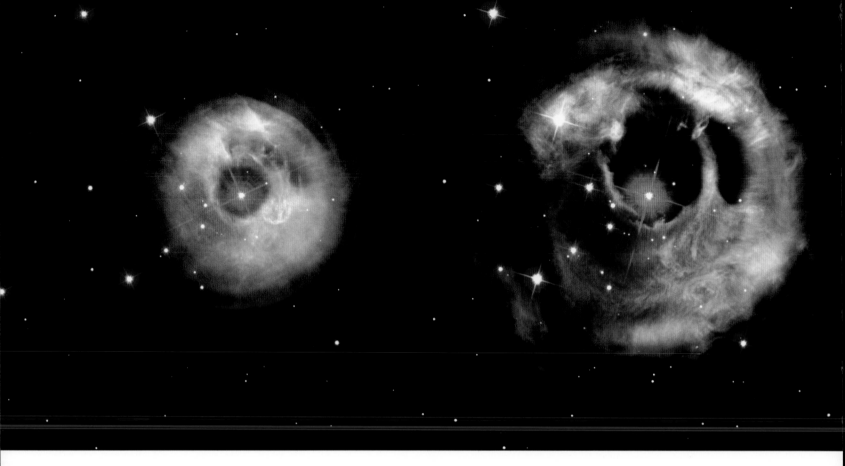

red giant Sun will span an angle of 60° across the skies of a totally scorched Earth.

During the red giant phase, a star loses most of its outer layers in shells that are blown away by radiation coming from the inner regions. The Hubble Space Telescope has recently uncovered the remarkable history of previous expansion shells around an ageing star called V838 Mon, which lies about 20,000 light years away. In January 2002 the star unexpectedly emitted vast amounts of light energy, which rapidly propagated into space like a photographic flash, along the way illuminating previously invis-

Previous pages:
This image of what is known as the Bubble Nebula reveals with remarkable clarity the ejection of hot gas from a star 40 times more massive than the Sun. The luminous central star is driving a substantial stellar wind moving at seven million km per hour (four million miles per hour). The ejected gas has formed a bubble six light years across, due to the action of the fast wind from the massive star ploughing into denser, slower moving material further ahead.

ible surrounding shells of dust that mark V838 Mon's earlier history. The rare and remarkable images reveal details of the structure of the shells as they expanded out to vast dimensions.

Eventually all Sun-like stars lose their entire outer envelopes of gas. This ejection leaves the stars with only the very core made of carbon, which will become compressed to about the size of the Earth. Known as a white dwarf star, the radiation from this initially very hot core lights up the expelled clouds of gas to produce a striking structure called a "planetary nebula". Planetary nebulae look like puffs or rings of smoke and signal the death of lightweight stars such as the Sun. They are, in effect, the ghostly remains of stars. Some of the most beautiful sights recorded in space have been provided by planetary nebulae. One particularly fine example, captured by the Hubble Space Telescope, being the Helix nebula. Lying about 650 light years toward the constellation of Aquarius, the nebula is seen as a colourful and intricate structure spanning almost three light years across, the red and blue glowing gases containing comet-like filaments that point toward the dying central hot white dwarf star. The

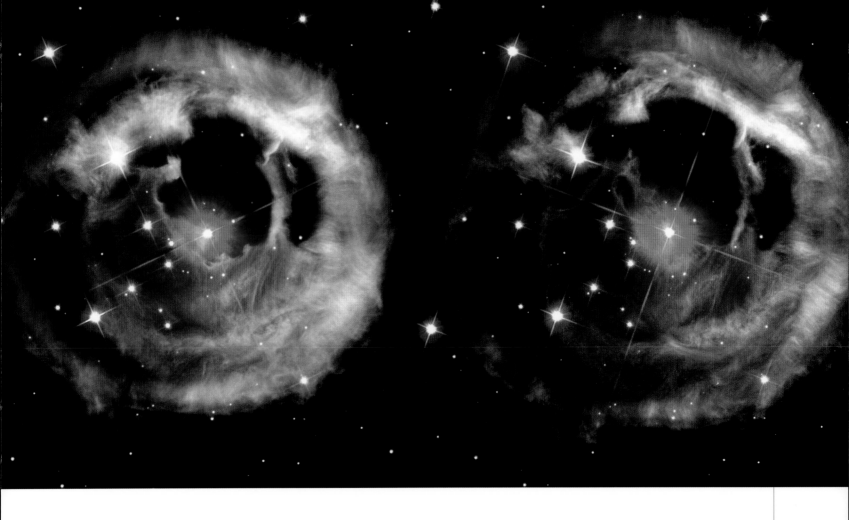

Helix has been sculpted by the complex interaction of fast winds of hot gas from the white dwarf, and previously ejected cooler shells of gas and dust. Remarkably detailed images such as these are allowing astronomers to unlock the mysteries of the final phases of stars in our Galaxy. A current challenge for astronomers is to understand why such a diverse variety of shapes and structures are seen in different planetary nebulae. A typical gallery would include almost perfectly circular nebulae, striking butterfly-shaped nebulae, some with pinwheels or spiral features, and others with overlapping concentric rings. One possibility is that the most non-circular examples are so shaped because the original dying star is actually one of a pair of stars, known as a "binary system".

More than 1500 planetary nebulae have been identified in our Milky Way Galaxy. On average, one new nebula comes into existence every year. Planetary nebulae present a superb laboratory for understanding the interaction between radiation and matter. The very hot central star emits radiation that is processed by the nebula, which contains solid, molecular, and atomic matter.

These stunning images from the Hubble Space Telescope track a series of previously ejected dust shells from an erupting star called V838 Mon. A burst of light from this bizarre star is spreading into space and reflecting off different surrounding shells. Over a time period of eight months, from left to right, the diameter of the structures illuminated spans from four to seven light years. V838 Mon itself is the red star at the centre of each image, located about 20,000 light years from Earth.

Out with a bang

Stars that are born with a mass of more than ten to a hundred times that of our Sun perish in a far more dramatic and violent manner. About one in a thousand stars in our Milky Way Galaxy is of this magnitude and uses up its resources of nuclear fuel for fusion at a much greater rate than the Sun, maintaining a phenomenal power output. During this accelerated evolution these enormous stars can undergo very violent and eruptive stages. Due to their very high mass, the intense radiation in the inner regions of the star can push outward onto the upper layers,

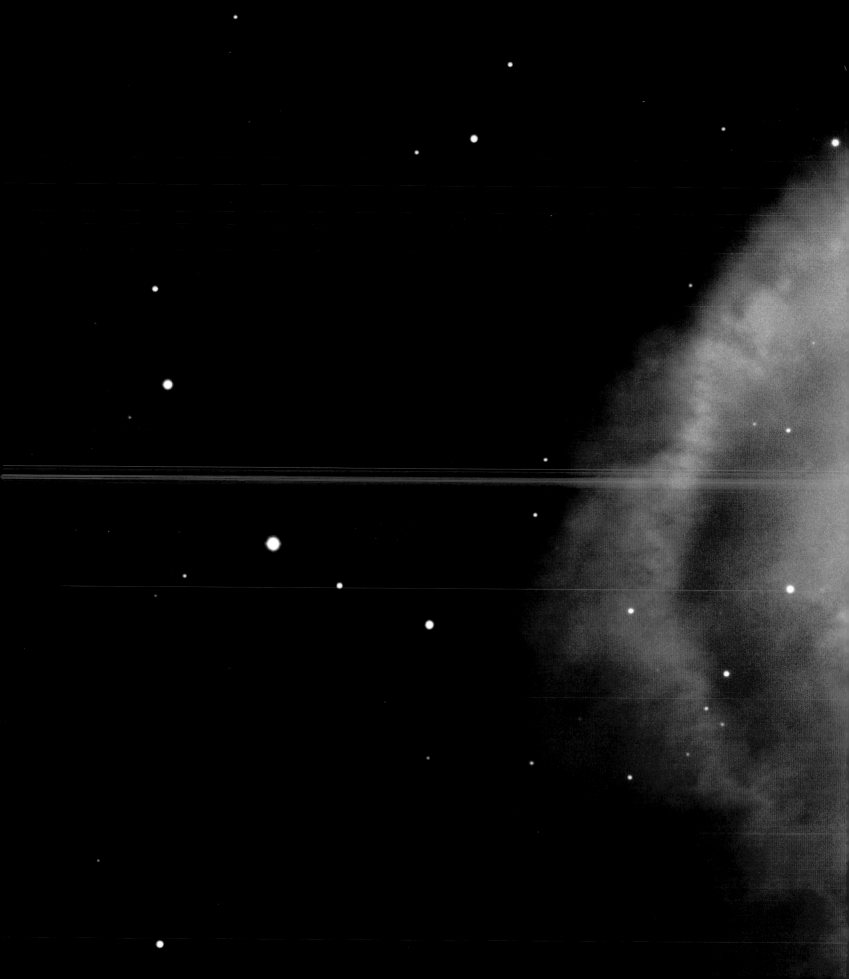

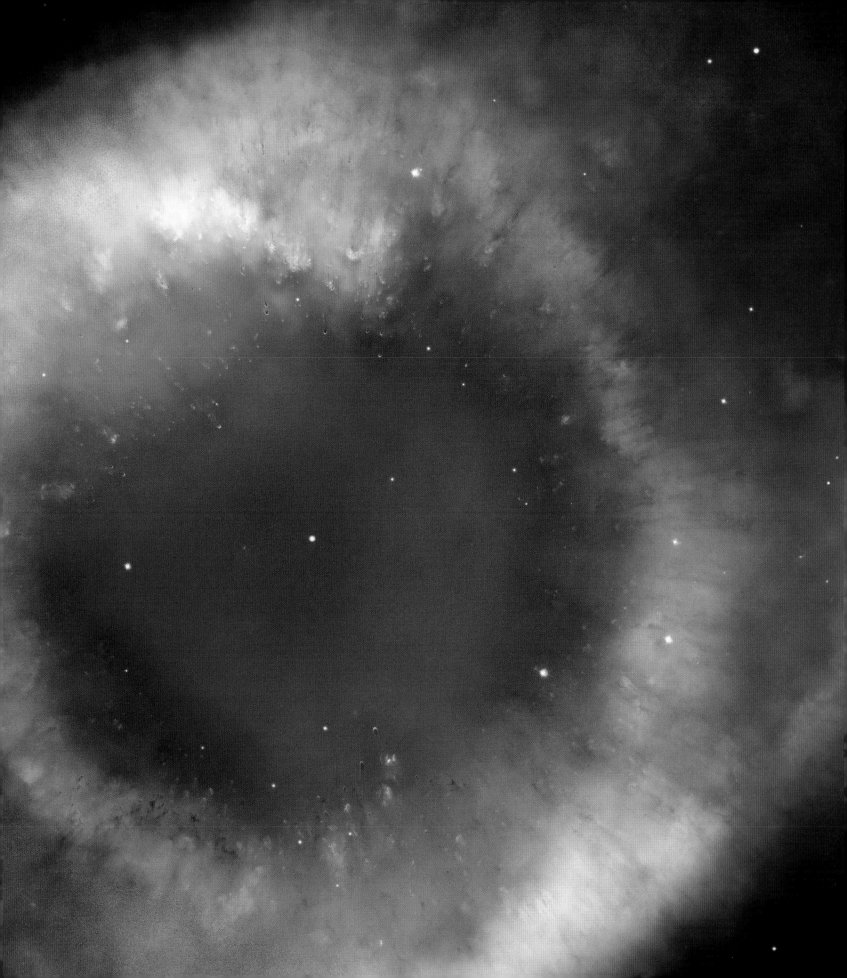

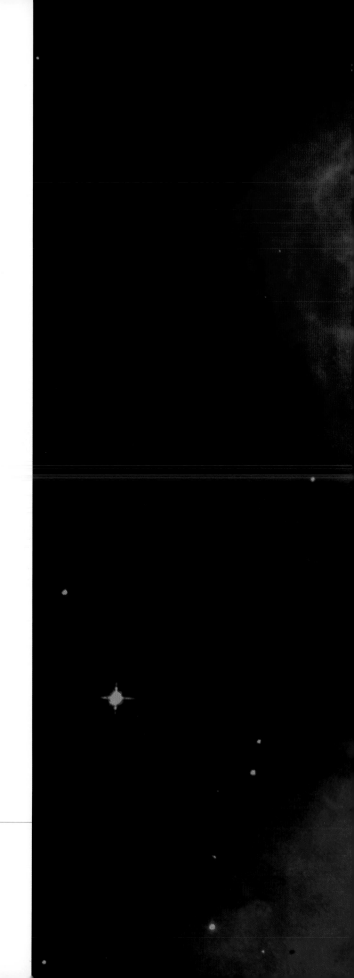

Previous pages: Intricate details of a planetary nebula known as the Helix are revealed by the sharp vision of the Hubble Space Telescope. The gases from this glowing, light-year-wide cloud were ejected toward the end of the life of a Sun-like star. The dying central star, known as a white dwarf, is seen at the centre of the nebula.

making the whole star highly unstable and forcing vast ejections of matter. In the case of the largest bodies, these giant outbursts can spew out ten times the entire mass of the Sun in a single episode. The display of fireworks seen around the massive star called Wolf-Rayet 124 is an awesome example of this transient behaviour. Huge blobs of gas are being expelled at speeds of more than 160,000km per hour (100,000 miles per hour), giving rise to a remarkably powerful stellar wind, and forming a clumpy and filamentary nebula around itself. Stars such as Wolf-Rayet 124 are highly unstable and heading for a most catastrophic end within only a few million years.

Massive stars can continue nuclear fusion reactions and extract energy from successively heavier elements such as carbon, oxygen, neon, silicon, and so on, until the stellar core is finally composed of iron. Iron is the most stable form of nuclear matter, and no energy can be extracted from fusing it to any heavier element. Once a central iron core is created, the "heavy-weight" star will no longer be able to generate heat to balance gravity, and destruction awaits. The star collapses very rapidly toward a supernova explosion, as billions of years of stellar evolution are undone within a few seconds. Travelling at 250 million km per hour (155 million miles per hour), the entire envelope of the massive star collapses onto the hard iron core, rebounding off it to create an outward travelling shock wave. This blast propels all the stellar material surrounding the core away into space, at speeds of more than 36 million km per hour (10,000 km per second, or 22 million miles per hour).

Supernovae are among the most powerful explosions to occur in the universe, each one releasing more than a hundred times the energy that the Sun will radiate over its entire ten-billion-year

A revealing overlay of X-ray (blue) and optical (red) images of the Crab Nebula, which is the remnant of a supernova explosion recorded by Chinese astronomers in AD 1054 . These superb images show the highly turbulent environment around a rapidly rotating neutron star. Bright wisps of gas are being launched away at half the speed of light.

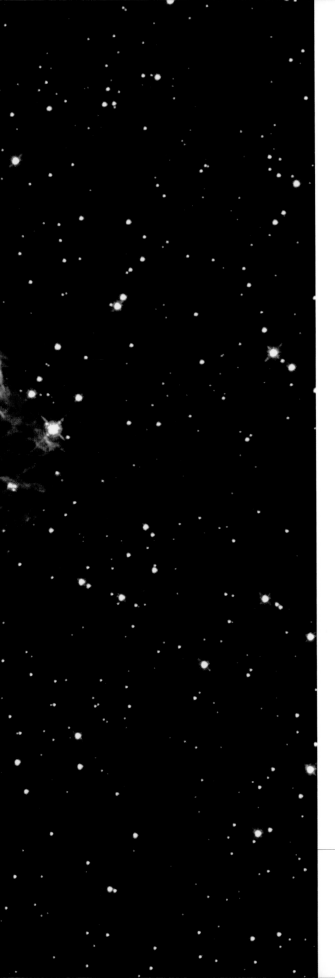

life. For several weeks a single supernova explosion can appear as bright as a galaxy of billions of stars. So what is left behind after this final act of stellar destruction? If the original (birth) mass of the star were ten to 20 times the mass of the Sun, then the iron core would be compressed into a bizarre ball of neutrons, barely 10km (six miles) across, forming a neutron star. These are sometimes detected from Earth as "pulsars", so named because they emit pulses of radio waves. If, however, the mass of the progenitor star were greater, then gravity scores the ultimate victory and the stellar core is collapsed to form a mysterious tombstone called a black hole. We will return to the strange properties of black holes later in this chapter.

All that remains is the stuff of life

The debris that remains after a supernova explosion takes the form of a hot and turbulent cloud called a "supernova remnant". New images from NASA's Chandra space observatory are providing valuable information about supernova remnants by studying their high-energy gas in the X-ray wave band and highlighting the very dynamic environment that surrounds the collapsed neutron stars. The collection of intertwined loops, filaments, and shells is composed of a rich variety of chemical elements, including oxygen, nitrogen, and iron. Initially produced inside the star as a result of nuclear fusion, these elements are vitally important for the development of known biological life. These life-giving chemical elements can only be created in the nuclear cores of high-mass stars. Ejecta from various ancient supernovae has been drifting through our Galaxy for billions of years, and seeded the giant molecular cloud out of which our own solar system formed more than five billion years ago. Our own bodies, and the Earth itself, are all composed of this stardust. As part of a cosmic recycling plant, the debris from supervovae explosions also provides new material for the next generation of stars.

The ghostly remains of gas and dust after the destruction in a supernova of a massive star, thousands of years ago. Known as N49, the debris from the violent event is about 150,000 light years away in the constellation of Dorado.

Colliding galaxies

We saw earlier in this book that unprecedented views of the very distant and young universe are revealing evidence of widespread collisions between dwarf galaxies. The frequency of collisions was higher during these early epochs than today simply because, in an ever expanding universe, everything was then closer together. It is becoming clear that these collisions have played a dominant role in the evolution of grand spiral and elliptical galaxies. Hubble Space Telescope images of young galaxies, barely a few billions years after the Big Bang, show them to be distorted and disfigured by the collisions, their shapes markedly different from the orderly elliptical and spiral galaxies we see around us today. Fortunately we can learn a great deal more about the mergers of galaxies by studying some nearby examples, which can be viewed in much finer detail.

Collisions between galaxies are phenomenal events involving tremendous forces; after all, the impacts involve two giant objects, each loaded with perhaps billions of stars, hurtling into each other at speeds of millions of kilometres per hour. Approximately one in a hundred galaxies seems to be undergoing such a collision, drawn together by the force of gravity over a period of hundreds of millions of years. Such an enormous timescale means of course that astronomers cannot watch an encounter take place in its entirety, but have to settle instead for snapshots of mergers and interactions at different stages of the process. The physical interaction between them most dominantly affects the free gas that exists in the medium between the stars; the impact generates a powerful shock wave that can drive through the gas clouds, compressing them until they are dense enough to set off the formation of new stars. By contrast the individual stars themselves are relatively unaffected because the average distance between them is several light years, millions of times greater than the size of a typical star. It is therefore highly unlikely that individual stars will collide, though their distribution and orbits within the galaxies may well change.

A variety of distortions

The outstanding images now being acquired by astronomers of collisions between galaxies have revealed an interesting variety of forms. The spread of morphologies seen after two (or more) galaxies collide reveals many properties, such as the masses of the galaxies, how compact they are, the velocities at which they move, and the angle of impact. Some galaxies may only experience a glancing blow, while others are involved in violent head-on collisions. In some cases the interaction can

A spectacular view from the giant Subaru telescope in Hawaii showing the irregularly shaped M82 galaxy. Lying 12 million light years from Earth, the red filamentary structures seen perpendicular to the main disc of the galaxy extend more than 10,000 light years in each direction. Most of this material is driven by a super-wind from the galaxy's central region, although M82 has also had several close encounters with its neighbouring galaxies.

also give rise to magnificent spiral arms such as those of our own Milky Way Galaxy. Indeed it seems that our Galaxy has had several ancient encounters, and is heading for major new ones in the future.

Bridges and tails

More often than not two interacting galaxies will undergo a glancing blow, rather than a direct collision. A pair of close galaxies, locked by their mutual gravitational forces and orbiting one another, can pass close enough to cause considerable damage and distortions. The resu

can be the formation of vast bridges and tails of gas, dust, and stars that stretch hundreds of light years into space. A beautiful example captured by the cameras of the Hubble Space Telescope is the encounter between the pair of galaxies known as NGC4676. Located 300 million light years away, the tussling galaxies have stripped away vast tails of stars and gas, and streams of material can be seen flowing for hundreds of light years away from them. The gravitational forces involved in this 160-million-year-old merger have also resulted in the creation of numerous clusters of new prolific stars, which show up in the images as blue patches of light.

Another tussle can be seen in the constellation of Canis Major, acted out between the spiral galaxies called NGC2207 and IC2163. NGC2207 is the larger and more massive component, whose tidal forces are distorting the shape of IC2163. The common hallmarks of galaxy encounters are clearly evident here, including great stretches of matter, sheets of highly shocked gas, lanes of stardust, and bursts of brand-new star formation. The severe contortion of IC2163 includes two tidal streams of gas and dust, and the unusual oval shape of its main disc of stars. The merger is relatively young, with the two galaxies probably reaching their closest points about 40 million years ago. NGC2207 and IC2163 are locked in this embrace forever and will continue to damage and disrupt each other, until they eventually merge to form a single massive galaxy billions of years from now.

Vast streamers of stars and gas stripped from galaxies are the tidal debris from the cosmic collisions. They can be viewed as experiments being conducted in nature's own crash test laboratory. The amount of matter in the streamers presents clues about the energy of the impact and the content of the galaxies. For example, a key factor that determines the resulting shape of galaxies in a collision is the amount of mass contained in vast "dark matter" halos that are thought to surround every galaxy.

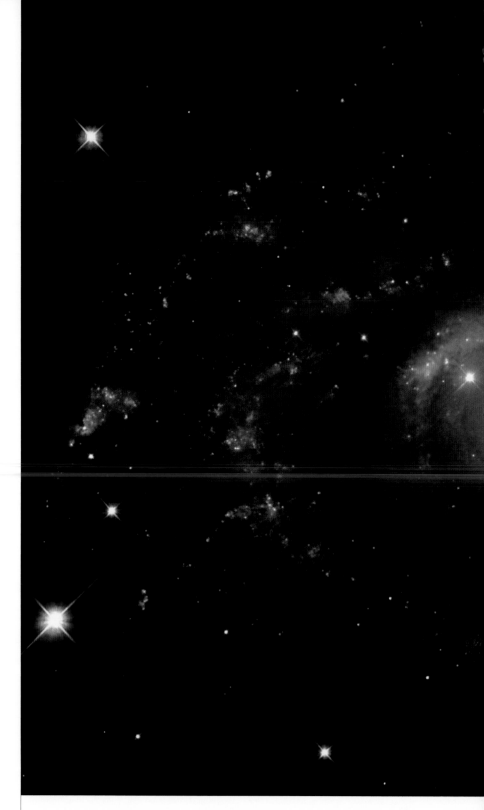

This majestic image captures the collision between two galaxies called NGC2207 (the large one on the left) and IC2163 (on the right). The massive gravitational forces at play are spewing out stars and gas into long plumes that stretch over 100,000 light years. IC2163 is swinging past NGC2207 in a counter clockwise direction, having made its closest approach some 40 million years ago.

Following Pages: Nicknamed the "Mice", the latest camera installed by astronauts during March 2002 on the Hubble Space Telescope witnessed this collision between a pair of galaxies. Vast streams of material are seen flowing from the distorted galaxies, with clumps of newly formed (blue) stars lighting up the "tails". Billions of years from now, the two galaxies will merge to form a single giant elliptical-shaped galaxy.

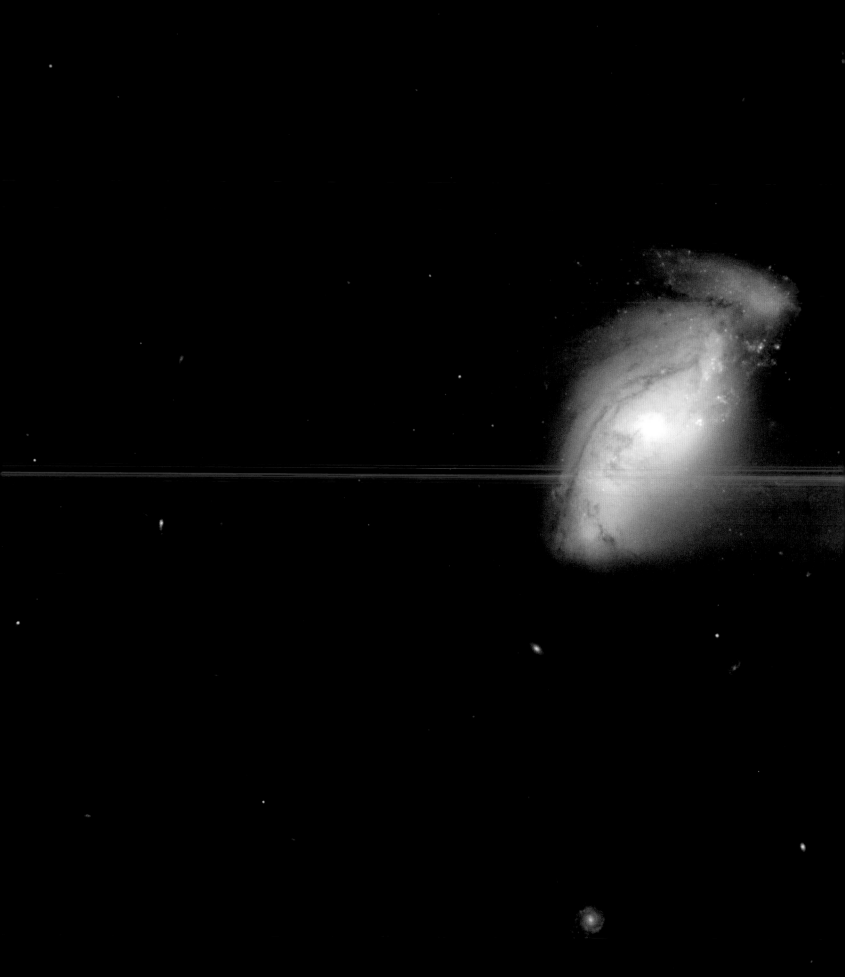

Rings, too

A direct, head-on collision between a small galaxy and a larger companion can sometimes result in a remarkable ring of stars forming around the main remnant galaxy. These "ring galaxies" signal an ancient bull's-eye impact. A fantastic case can be found about 600 million light years away in the constellation of Serpens. Known as Hoag's object, an almost perfect ring of blue hot stars is seen around a smaller central core of older yellow stars. Hoag's object is enormous; the disc of our Milky Way Galaxy would fit inside the span of the outer ring in Hoag's object, which is about 170,000 light years across. Astronomers believe that a smaller galaxy collided into the nucleus of the main galaxy about three billion years ago, causing a phenomenal wave of energy that travelled outward at terrific speeds. Much like ripples from stones thrown into a pond appear to merge and spread as they proceed from the centre, the wave compressed gas and dust as it progressed, becoming so greatly heated that eventually new star formation was triggered. These resulting stars are the young massive stars seen in the ring of Hoag's object. In this case there remains no sign of the impacting galaxy, and it seems likely that some of its debris has also been collected in the ring. Dozens of collision-formed ring galaxies have now been identified by astronomers, with numerous suspected examples waiting to be studied.

Compact groups

Galaxies rarely occur as isolated objects in space. They most often form groups of galaxies, and the collections can range from just a few members to several thousand, held together by the mutual pull of their gravity. When a dozen or so galaxies are gathered in such a small area that they end up touching and interacting with each other, they are known as a "compact group", and more than 300 examples have so far been discovered. Their significance for astronomers is that they serve as nearby examples of the types of merging activity thought to have

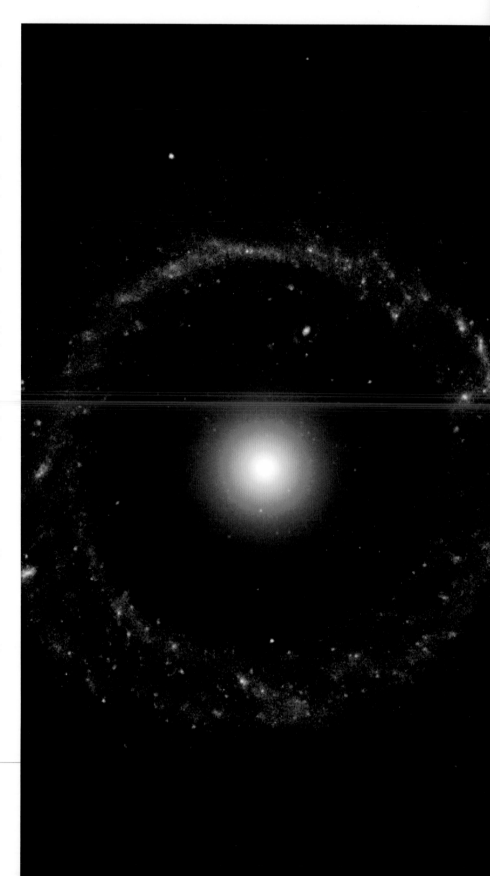

A stunning ring-shaped galaxy known as Hoag's object is viewed here. The galaxy is about 120,000 light years across, at a distance of 600million light years from Earth. The blue ring of young stars may be the disrupted remains of an ancient collision with another galaxy.

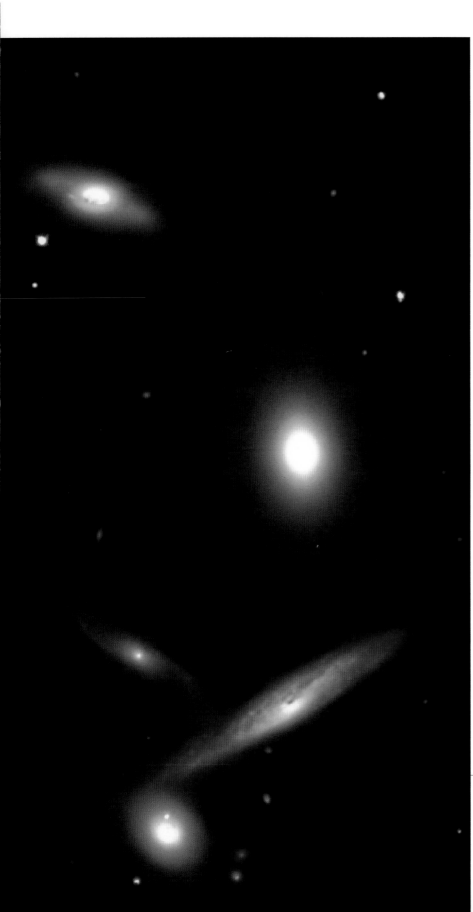

taken place in the early history of the universe, when galaxy encounters were very common. The compact groups are comparatively close and so can be viewed in superb detail to understand the havoc triggered by the collisions.

The compact galaxy group called Hickson 40, lying 300 million light years away toward the constellation of Hydra, is made up of five spiral and elliptical galaxies that appear to be physically interacting. Distorted by awesome gravitational forces, huge streamers of gas and stars are being spread over vast distances, in the beginnings of the eventual merger of some members of the group. Subsequent squeezing of this gas can trigger the formation of brand-new stars. Studies of Hickson 40 are providing fresh perspectives on how powerful encounters shape the evolution of individual galaxies.

Galactic cannibalism

In the violence of the universe, the ultimate encounter is when a galaxy is completely swallowed by one of its much larger neighbours, resulting in its complete disruption. These acts of galactic cannibalism are often found to occur in the central regions of groups of galaxies that are rich in members. Here, the core galaxy can consume several dwarf ones to grow into a super-massive galaxy, five to ten times the size of our own Milky Way. The impressive spiral galaxy ESO 510-G13 is an example which is apparently in the process of devouring its smaller companion. Viewed edge-on, almost 150 million light years from Earth, the galaxy has an unusual twisted and warped disc of stars caused by this major disturbance. The contortions driven by the gravitational forces have also led to the formation of young blue stars in the outer regions of the galaxy. The complete merger of ESO 510-G13 with its neighbour into a single larger galaxy will be completed over the course of a few million years. Understanding how this encounter proceeds can provide astronomers with new insights into the structure and stability of the original larger spiral galaxy, such as how densely it was packed and how material was spread from the core to the outer regions.

This image from the Subaru telescope shows a compact group of galaxies known as Hickson 40. The galaxies are so tightly grouped that the three members toward the bottom of the image are touching each other and interacting under the action of gravity.

The fate of our Milky Way Galaxy

Our own spiral-shaped Milky Way Galaxy is part of an ensemble of 45 galaxies known as the Local Group that are all contained within an area about three million light years across. There are only three large spiral galaxies in this group, including the magnificent Andromeda, the rest being comprised of small ("dwarf") elliptical galaxies and irregular-shaped ones. Their combined gravity binds this Local Group together, and there is plenty of evidence to suggest that our Galaxy has also been the victim of collisions with the smaller members. We can ask then, what is the likely fate of our Galaxy in this interacting group?

One way that astronomers attempt to answer this question is by simulating the merger of galaxies using computer models. Deriving some basic constraints from observations of the speeds and masses of the galaxies, high-powered computers can then produce animated sequences displaying the evolution of a collision that in reality would take place over a period of some 500 million years. Observations of our Local Group of galaxies have shown that our own Milky Way Galaxy and the twice as large Andromeda Galaxy (also called M31) are in fact heading toward each other at 480,000km per hour (300,000 miles per hour). They are separated by just over two million light years, and travelling at these speeds they are due to collide about five billion years from now. The current prediction is that the encounter will initially be a glancing sideways impact, rather than a full-blown head-on one. Nevertheless, the computer models suggest drastic changes will follow. After the first blow, the two galaxies will be bound into an ever decreasing orbit, with subsequent second and third major impacts. Eventually, over roughly a billion years or so, the two galaxies will be drawn closer and closer together until they merge into a single giant elliptical galaxy. How will this affect us on Earth? It turns out that future inhabitants of our planet will have a more local crisis to deal with much earlier than this full merger: the Sun will by this time have progressed along its destined evolutionary path and bloated into a red giant star, leaving a scorched, lifeless Earth with a burnt atmosphere and boiling oceans.

The collision between two galaxies is simulated here using calculations performed on powerful computers. From left to right (and on following pages), the visualizations provide insights into the billion-year period over which two spiral galaxies are pulled together by their mutual gravity, with the subsequent formation of vast tidal tails or streamers.

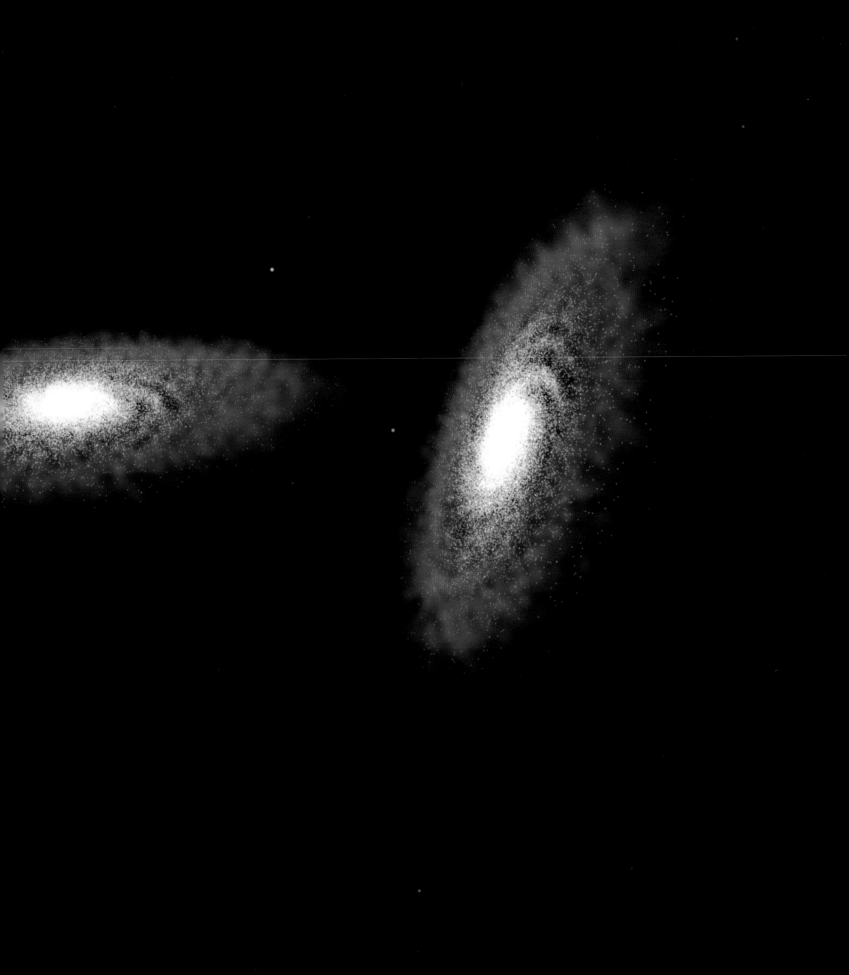

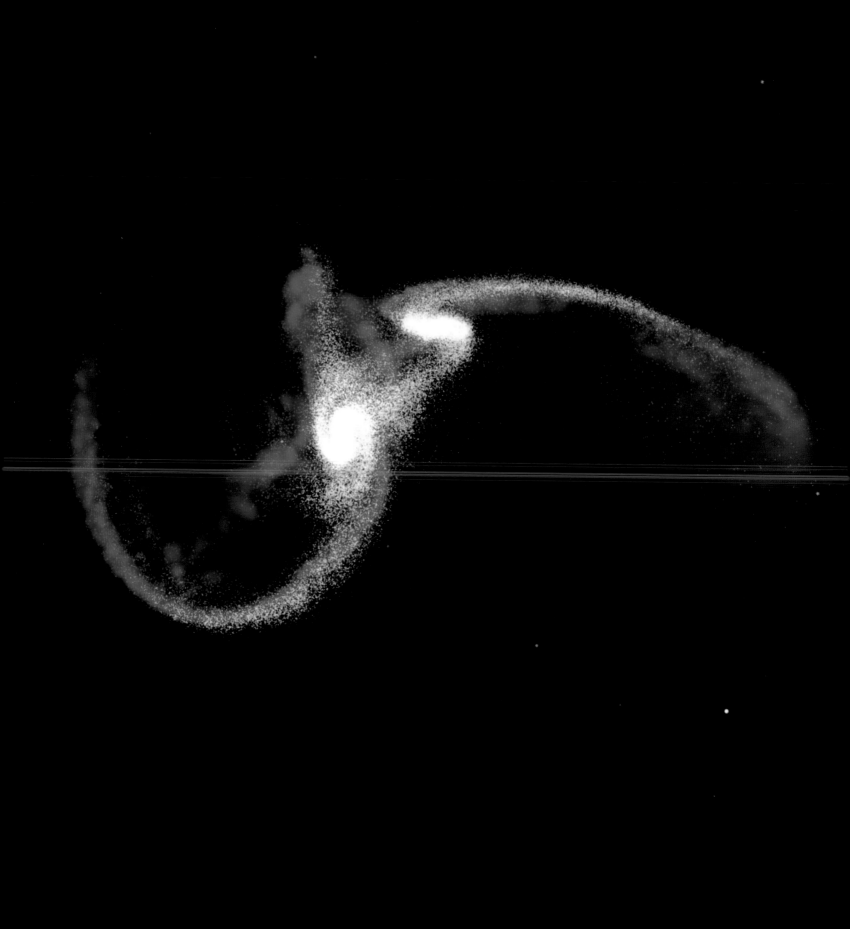

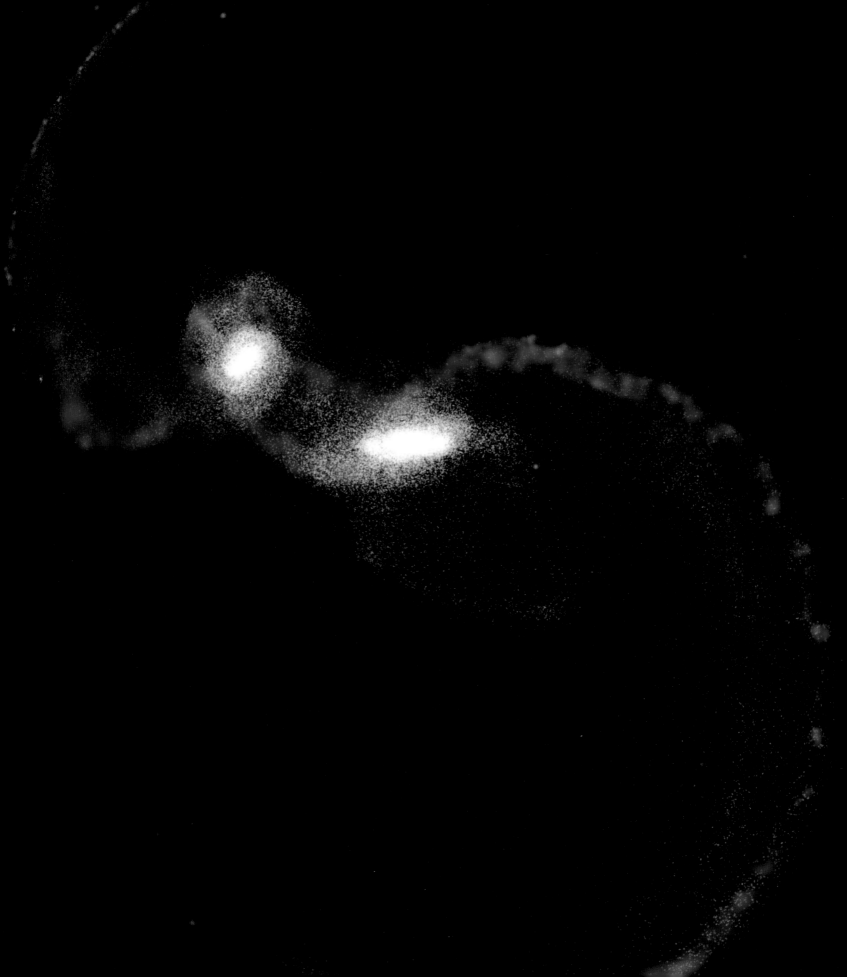

What lurks at the centre of our Galaxy?

In recent years astronomers have come to realize that the region around the centre of our Milky Way Galaxy is a tremendous and unique laboratory, where a large number of very important and energetic phenomena are to be found, occurring on many different scales. The study of these processes can be very challenging and often demands data from the most powerful telescopes operating in several wavebands, from X-rays to radio. The greatest challenge of all is in unveiling the monstrous object that lurks at the centre of our Galaxy.

Let us start with an overview of the basic structure of our Galaxy. We live inside an average-sized spiral galaxy that contains about 150 billion stars. Like all spiral galaxies, our Milky Way consists of three main components called the disc (which is where the solar system resides), a central bulge, and a vast diffuse halo that surrounds the entire disc. The disc contains the defining spiral arms, which are strips of gas, dust, and stars that unwind out of the central bulge. Our Sun resides inside one of these arms, called the Orion arm. As we are within the disc, we actually have a very poor overall view of our Galaxy; it is like trying to work out the shape of a large forest while standing deep among its tall trees. The disc of our Galaxy is about 100,000 light years across, but it is comparatively slim, with a thickness of only about 900 light years. The disc is surrounded by an extended spherical region, or halo, that is more than 150,000 light years in diameter. This halo contains the oldest stars in the galaxy and is also host to significant amounts of enigmatic dark matter. More than half

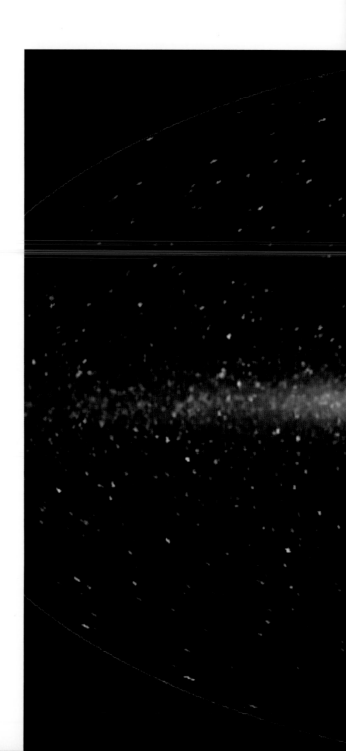

An infrared image of our Milky Way Galaxy taken by the COBE satellite. The edge-on view of the thin disc of our Galaxy is clearly revealed, along with the Galaxy's central bulge, which contains a dense concentration of stars, gas, and dust. The Galactic disc may be seen across very dark and clear skies as a milky band of light.

the mass of our Galaxy is contained within the central bulge, which has dimensions of about 10,000 light years in width by 1000 light years in height. The very centre of our Galaxy lies at the centre of this bulge of densely packed stars and gas. The entire spiral galaxy rotates around the centre, with increasing speed as you move outward along the disc. Our solar system has a rather mediocre location, situated about 28,000 light years from the centre, where it rotates at about 900,000km per hour (550,000 miles per hour).

Toward the turbulent centre

Viewed from Earth, the centre of our Galaxy lies toward the constellation of Sagittarius. We cannot, however, just take a large telescope and peer into it, as our view is very heavily obscured by incredibly dense concentrations of gas and dust. Toward the centre of our Galaxy the incidence of stars is a million times greater than in the neighbourhood of our Sun; if we lived within the central region, the brightness and numbers of stars would be so great that we would live in conditions of perpetual daylight.

To study the central regions of our Galaxy, astronomers use telescopes that work at infrared wavelengths, as these render veils of dust increasingly transparent. Approaching within a few hundred light years of the centre, the first surprise is the discovery of thousands of massive and highly luminous stars, some of which are hundreds of times the mass of the Sun and factors of a billion times more powerful. They are embedded within very prolific regions of star formation that represent sources of enormous energies. It is thought that these massive star clusters were created almost simultaneously in a phenomenal burst of star formation, during the early stages in the assembly of our Galaxy, and have a frantically short life cycle. Indeed the central regions of the galaxy are littered with numerous supernova remnants.

High-quality images taken at radio wavelengths also allow us to look more deeply into the regions surrounding the very centre of the galaxy, known as Sagittarius A, where traces can be seen of the violent motion of vast threads and streams of material. The central hundred light years or so are heavily scarred by the turbulent environment of swirling clouds of gas, powerful radio beacons, and tight clusters of millions of stars. Shell-like structures and thousands of strong X-ray emitting sources are also embedded here, all testament to the high rates of supernova explosions. All this activity betrays an awesome object at the very centre, which is capable of generating tremendous power.

Previous pages: The view toward the centre of our Galaxy in the direction of the constellation of Sagittarius. Tuned to infrared light, this astounding image from the 2MASS space mission reveals a remarkable ten million stars in this segment alone, which spans a patch in the night sky equivalent to an adult fist held at arm's length.

This substantial cluster of stars located toward the very centre of our Galaxy contains a mixture of young stars and highly unstable luminous stars heading for destruction as supernovae. The cluster is also home to one of the most powerful and massive stars in our Galaxy, named the Pistol Star.

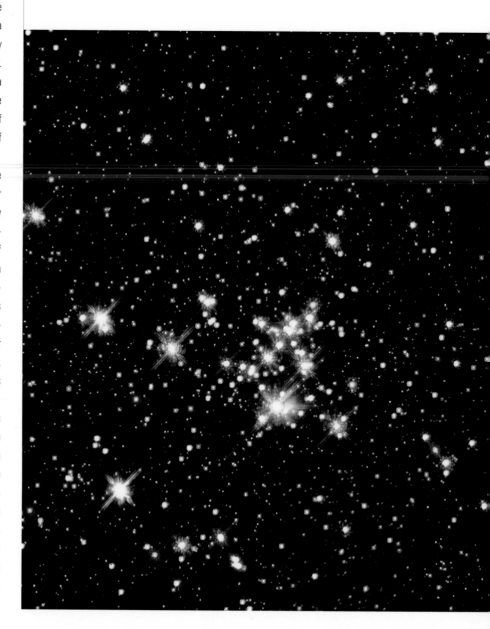

Homing in on the "beast"

To reveal the enigma of the Galactic centre, astronomers have been making precise measurements of the motion of individual stars that lie within just one light year of the very centre. This is a very demanding task because it is difficult to detect pinpricks of starlight through the obscuring dust clouds. Using giant instruments such as the Keck 10m telescope in Hawaii, and the European Southern Observatory's NTT telescope at La Silla in Chile, it has recently been possible to record several stars as they orbit tightly around the centre. The speed of these stars can yield information about the mass of the object they are circling, in the same way that we can determine the mass of the Earth from the motion of our Moon. It turns out that the stars are moving at speeds of up to five million km per hour (three million miles per hour): as the time taken for them to complete an orbit is only a few decades, they must be circling a very massive object indeed.

One star, labelled S2, is particularly revealing. Using data gathered over a period of ten years, it has just become possible to map S2's entire oval-shaped orbit, revealing how its speed increases as it comes closest to the Galactic centre, and then decreases as it travels further away. Using these results astronomers have calculated that the central object pulling S2 around must have a mass of about three million times that of the Sun. As all this enormous mass has to be contained within a fraction of a light year, we now have compelling evidence that the object lurking at the centre of our Galaxy is a super-massive black hole. It is the equivalent of squeezing almost three million stars like the Sun into a space barely the size of our solar system: nothing apart from a black hole could account for such a high density of material.

NASA's Chandra X-ray satellite has recently secured the deepest yet X-ray images of this super-massive black hole. They have revealed mysterious short-lived bursts of X-rays arising from close to the point of no return around the black hole. Huge lobes loaded with gas that has been heated to more than 20 million°C (65 million°F) can be seen, providing clues as to the manner in which the black hole is being fed with material.

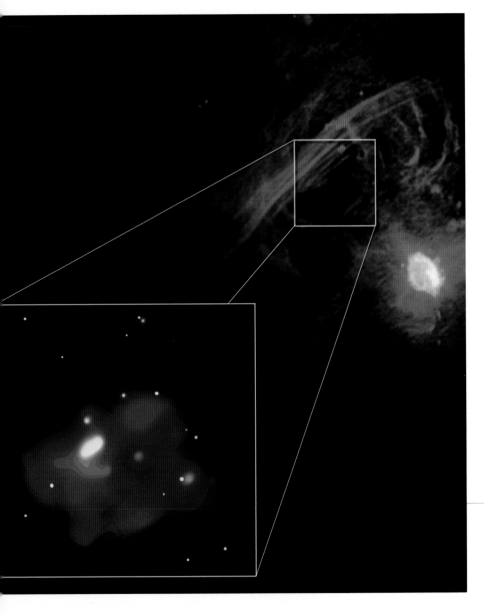

The violent and turbulent region surrounding the Galactic centre is highlighted in this radio-wavelength (red) image, which also contains filaments of million-degree hot gas witnessed in the X-rays (blue image). The X-rays are produced when high-speed electrons collide with clouds of surrounding cold gas.

Probing the black hole

Black holes are very bizarre objects where the gravitational field is so strong, and the space around them so heavily distorted, that nothing, including light, can escape from them. There is a very critical boundary between the inside of a black hole and the rest of the universe outside, known as the "event horizon". Nothing ever gets out from within the event horizon: it can be thought of as a one-way membrane through which matter and energy can fall into the black hole, but which allows nothing to escape in the opposite direction – material that crosses it is consumed forever. The larger the mass of the black hole, the larger the radius of the event horizon. In the case of the three-million solar mass black hole at the centre of our Galaxy, this radius is only about the equivalent of 50 times the distance between the Earth and the Sun, which is a remarkably tiny dimension on the overall scale of a galaxy.

The stormy radio and X-ray emissions detected coming from Sagittarius A are actually beamed by material as it falls toward the event horizon, but before the boundary is crossed. In-falling gas is heated as it swirls around the massive black hole, and some of it is then blasted away before it gets beyond the point of no return. These violent outflows and giant spurts of material may actually be starving the black hole of material by reducing the amount of matter it can suck in. Indeed the black hole at the centre of our Galaxy may sound impressive, weighing in at three million times the mass of the Sun, but it is comparatively puny: we will see in the next section that some galaxies host black holes that are a thousand times more massive.

While it is widely believed that black holes are to be found at the centres of almost all substantial galaxies, there is still debate as to how super-massive examples such as that in our own Milky Way originated. One possibility is that a central black hole, originally arising from the collapse of multiple massive stars, grows steadily as matter continually flows into it after being shed by surrounding stars. Another theory holds that a central black hole could gain mass as a result of the merger of the cores of two smaller galaxies after a direct collision.

A superb Chandra satellite X-ray image of our Galaxy's central region, surrounding the object known as Sagittarius A . The intensity of more than 200 X-ray emitting sources seen in the region has been used to constrain the mass of the super-massive black hole at the core of our Galaxy.

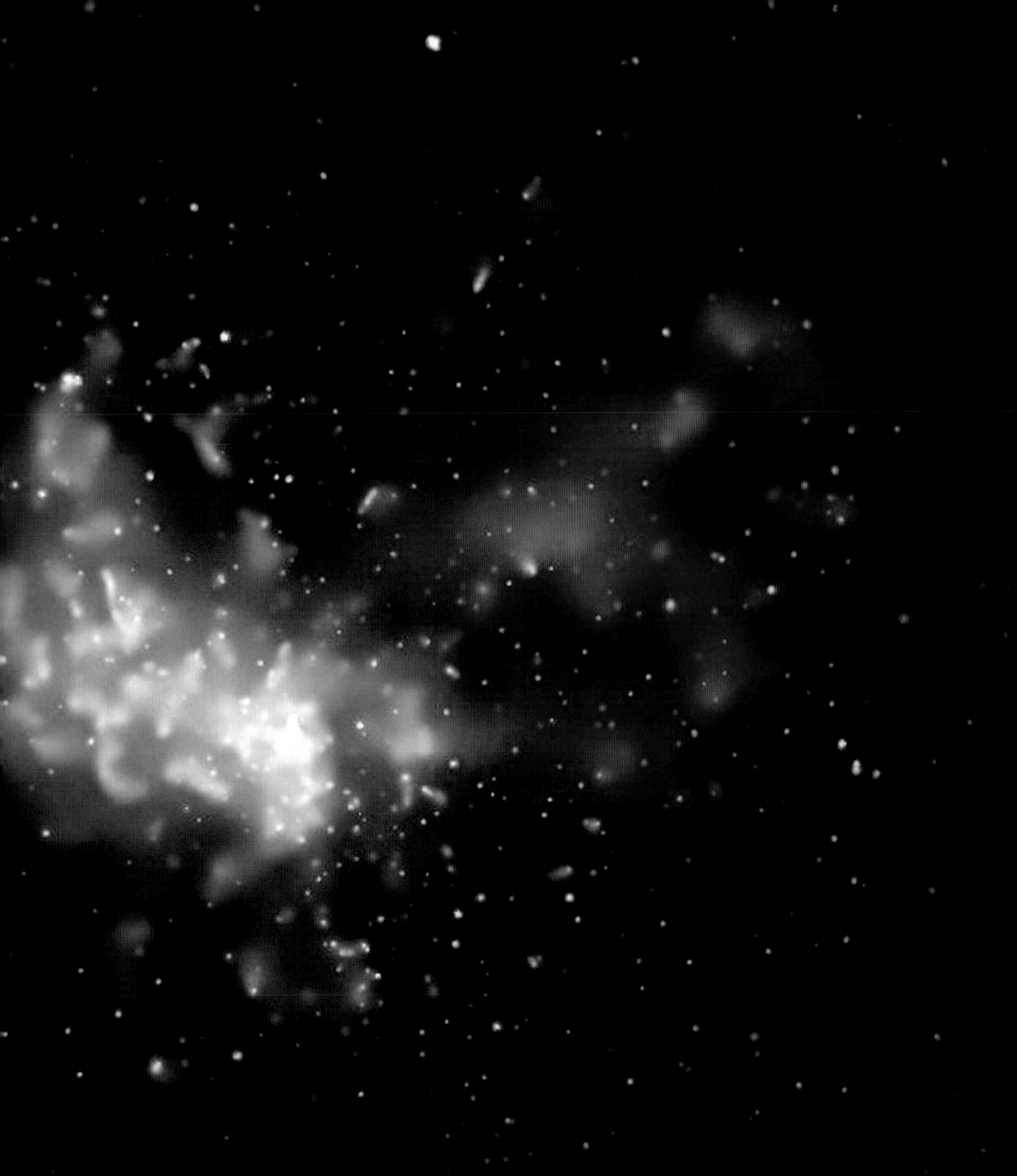

The power of quasars

Quasars are the most luminous, powerful, and energetic objects known in the universe. They inhabit the very centres of active young galaxies and can emit up to a thousand times the energy of our entire Milky Way. This intense emission is detectable over a wide range of the electromagnetic spectrum, from low-energy radio waves to high-energy gamma rays. What is remarkable about this process is that this prodigious amount of energy is being generated within a region that is not much bigger than our solar system. It is akin to the bulb from a small torch emitting the same amount of light as all the buildings in an entire city.

It is now widely accepted that the engine which drives this phenomenon by providing such immense power is a rotating super-massive black hole at the centre of the quasar. Measurements of the amount of radiation emitted by the quasar suggest that the mass of the remarkable black hole is typically one billion times the mass of the Sun. Quasars are so luminous that they can be seen over great distances. The most distant quasars are more than twelve billion light years away, which means we are viewing them at a time when the universe was less than a tenth of its current age. Quasars are therefore a valuable resource in understanding the way luminous matter is distributed in the early universe, and act as powerful searchlights that can illuminate matter residing in the vast, dark space between galaxies. Quasars are also the most powerful known sources of X-rays, and studying this type of radiation, using modern technology such as telescopes mounted on spacecraft, is providing astronomers with new insights into the highly turbulent environments that surround quasars' central black holes.

As a quasar is such a luminous and outshining part of a galaxy, it is usually very difficult to discern what the overall host galaxy looks like. Much like the way the Sun's outer corona can be observed only when it is eclipsed by the Moon, the regions surrounding a quasar may only be viewed properly by attempting to block the light from the core, using special instruments called "coronagraphs". Using the "Advanced Camera" on the Hubble Space Telescope, the light from a well-known quasar called 3C273 was eclipsed using an on-board coronagraph. The host galaxy that was revealed was found to be more complex than had been thought previously. Findings showed that amid some traces of spiral-shaped structures like the arms in our Galaxy were lanes of dark dust and knots of hot gas being blasted away in rapid jets.

Recent images reveal a wide assortment in size, brightness, and morphology of the quasar hot galaxies. In several cases the galaxies appear to have highly compact, close companions, thus increasing the likelihood of violent interactions. The complex diversity of galaxy types raises the possibility of a variety of mechanisms for "switching on" the central quasar.

This stunning image of the highly active galaxy Centaurus A is a composite of X-ray, optical, and radio images. Jets of high energy particles are blasting away from a central super-massive black hole. The remarkable galaxy also hosts a ring of million-degree hot gas that spans 25,000 light years.

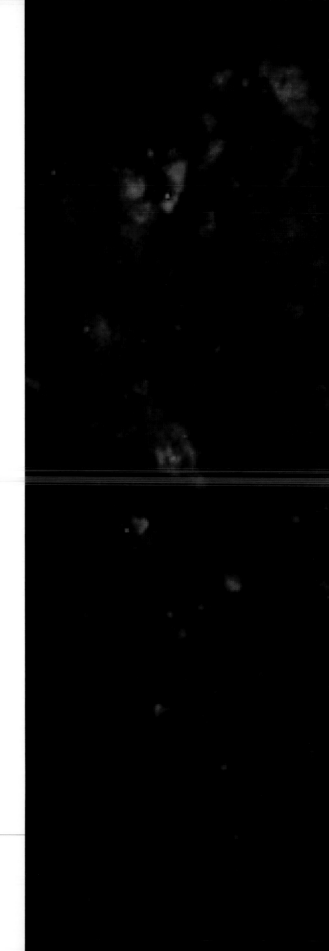

Feeding the black hole

The luminous power of a quasar depends on how massive the central black hole is and the rate at which it can consume matter; as the light emitted is so prolific, large quantities of gas must be pouring into the super-massive black hole. It is typically gorging on a steady diet equivalent to the consumption of several Suns per year. So what are the feeding and eating manners of quasars? The super-massive central engine is surrounded by a plate-shaped region called an "accretion disc". This disc is an active cauldron of swirling gas through which matter is spiralling into the black hole, like water down a plug hole. The accretion disc is typically barely a few times the size of our solar system, which is minuscule compared to the overall dimensions of galaxies. As the material from stars, gas, and dust spirals ever closer to the quasar's black hole, it spins faster and becomes more compressed. The effect is to violently rub the particles of matter against each other, creating frictional heat that raises the gas temperature to millions of degrees. In the innermost regions of the disc, the very hot matter radiates enormous amounts of energy, which is converted into the quasar's prodigious X-rays, radio waves, and optical light that are detected on Earth.

High-speed jets

Remarkably, as matter is being sucked into a quasar's super-massive black hole through the whirlpool-like accretion disc, some of the gas is also blasted away in very energetic high-speed jets. These jets, which travel out of the galaxy at almost the speed of light in very narrow beams, have been viewed stretching over several tens of light years. Scientists are trying to learn how matter is violently ejected from the core of the quasar and how the jets interact with the surrounding environment.

The tremendous debris pushed out at fantastic speeds contains very hot gas that shines in X-rays, and researchers using NASA's Chandra X-ray observatory have been able to reveal the explosive nature of these jets. For example, the jet forced from the quasar called 3C273 is not a smooth beam, but instead con-

The highly active nucleus of a galaxy called NGC1068 is uniquely revealed in this X-ray image from NASA's Chandra satellite. Incredibly hot gas can be seen blowing away at a fierce 1.5 million km per hour (1 million miles per hour) from the vicinity of a central super-massive black hole.

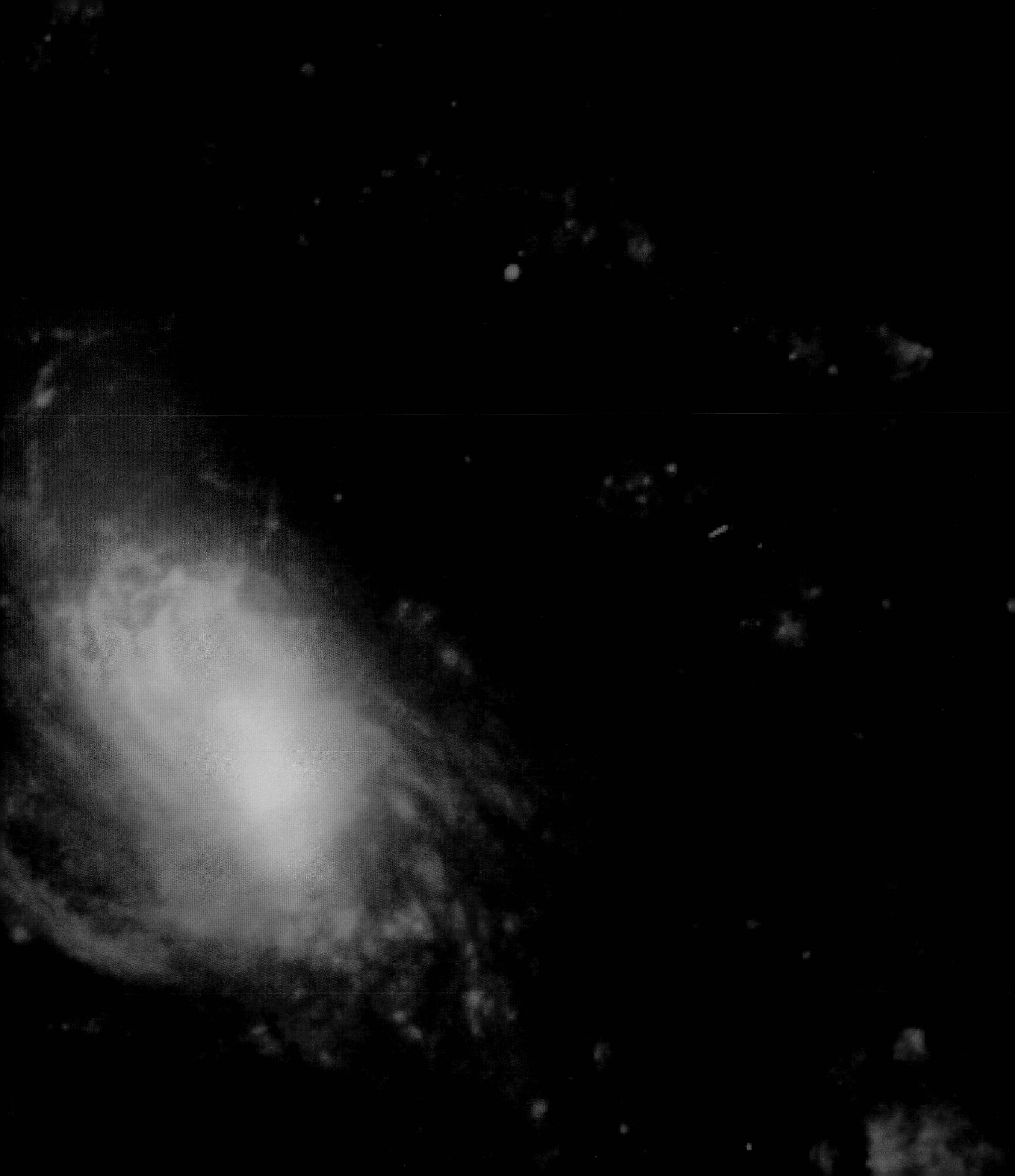

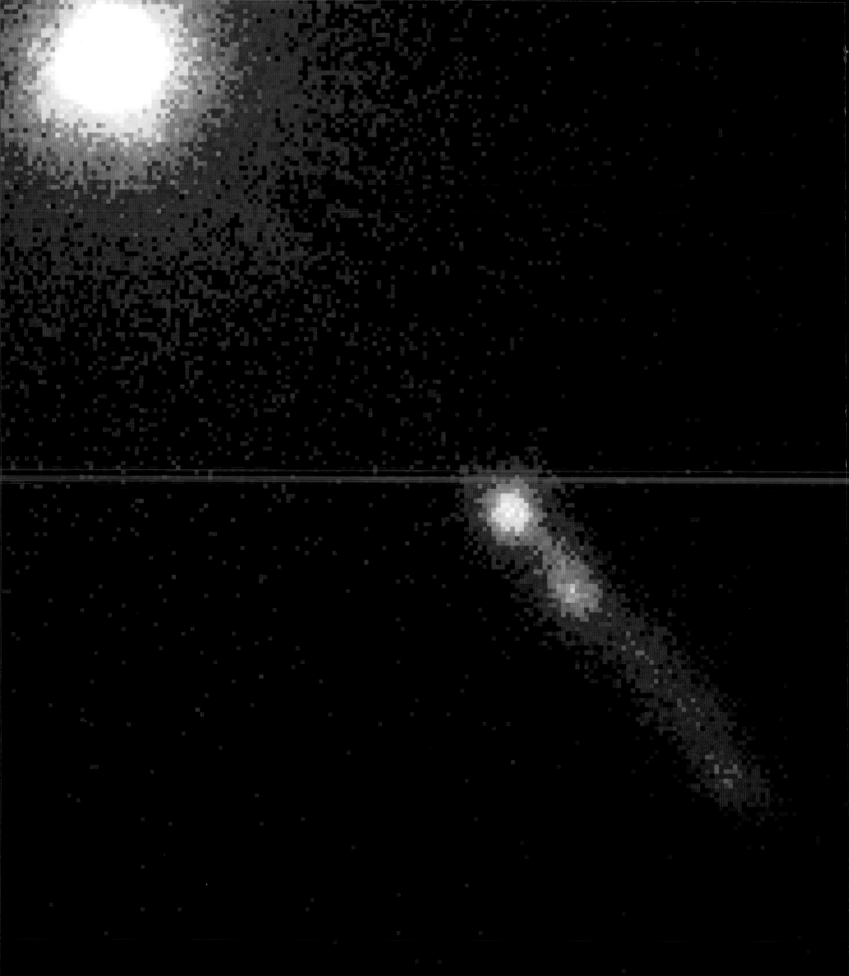

tains clumps of gas. The length of the jet and the knots of X-ray emission point to intermittent and variable activity very close to the central super-massive black hole. The jet originates from gas which is under tremendous pressure as it rapidly orbits in the whirling disc around the black hole.

The combination of in-falling material, rapid turbulent motion, and extreme energies can give rise to other bizarre and complex outflows, which are seen moving away from the quasar. The powerhouse residing in the galaxy NGC4438 has been captured by the Hubble Space Telescope blowing vast bubbles of hot gas into space. Located 50 million light years from Earth, material is being spewed out as part of the eating habits of the massive black hole. Already a staggering 800 light years across, the bubbles of gas will continue to expand until they eventually break up to become part of the environment comprising the outer regions of NGC4438. Another example has been viewed by the Chandra X-ray satellite, revealing a fast outflow rising from the super-massive black hole at the centre of the galaxy NGC1068. A vast cloud of gas is seen surrounded by a doughnut-shaped region of cooler material. The high speed wind has a temperature of about 10,000°C (50,000°F) and is propagating at 1.6 million km per hour (1 million miles per hour).

A powerful jet is shooting away from the quasar 3C273. Launched at very close to the speed of light, these new X-ray images demonstrate the lumpy and fragmented nature of the jets, which were previously thought to be a smooth-flowing stream of gaseous material.

The overall picture being built up of a quasar is of a hive of extreme activity confined to the central regions of a young galaxy. The driving engine for all the power, radiation, and blasts of matter is a black hole that has a mass equivalent to an entire normal galaxy, but with a dimension comparable only to the size of our solar system. This "monster" is fed through a rapidly rotating disc of gas and dust. Moving further away from this core is an ensemble of cooler gas clouds, travelling at lower speeds. Surrounding this system of clouds is a doughnut-shaped torus of even cooler gas and dust, typically about a hundred light years across. As the torus will look different if viewed edge-on or directly from above, a quasar will present different properties and characteristics depending on the angle of its orientation to us in the sky.

Life time of a quasar?

We have seen that quasars are capable of a tremendous rate of energy production, but it is generally believed that this prolific output can only be maintained for a short phase in the life of a galaxy. A direct consequence of the jets, bubbles, and other outflows discussed here is that they deplete the accretion disc of material close to the black hole. The expectation is that eventually a cavity is created in the rotating disc which effectively cuts off the food supply to the black hole, therefore reducing its energy generation rate. For the rest of its life the activity at the core of the galaxy would be much less than during its youth. Further periods of enhanced activity and outbursts are still possible, if, for example, the quasar galaxy collides and merges with another galaxy in such a way as to start loading the accretion disc

with material again. In other words, galaxy collisions can in effect restart a quasar.

The fact that all quasars reside at vast distances from us supports the idea that they represent the earliest stages in the life of a galaxy. It may well be that all galaxies go through a very active period in their youth, powered by massive central black holes (although they may not all host powerful quasars.) When over a period of billions of years their black holes are starved of fuel, the galaxies turn into relatively quiet, normal galaxies like the grand spiral and elliptical structures we see relatively close to us today. Their once great central powerhouses are now comparatively dormant, with only a shadow of their youthful activity. Our own Milky Way Galaxy was probably never a quasar-hosting galaxy. It most likely had an energetic youth, but, as noted earlier, it has a relatively modest black hole of about three million times the mass of the Sun, and not the billion solar mass object needed to have once been a quasar.

The observation of quasars at vast distances contrasts with their relative dearth nearby, suggesting that they were much more common in the early universe. This early abundance is consistent with the notion that the quasar "switches off" when the black hole has consumed all available fuel, as more mass is expected to be freely accessible in the early universe. It is clear that the formation and evolution of quasars, super-massive black holes, and galaxies are closely connected.

A super-massive black hole resides in the central core of the active galaxy NGC4438, which is located 50 million light years from Earth in the constellation of Virgo. This detailed image from the Hubble Space Telescope highlights huge bubbles of hot gas emanating from the highly turbulent hot disc of material surrounding the ravenous central object.

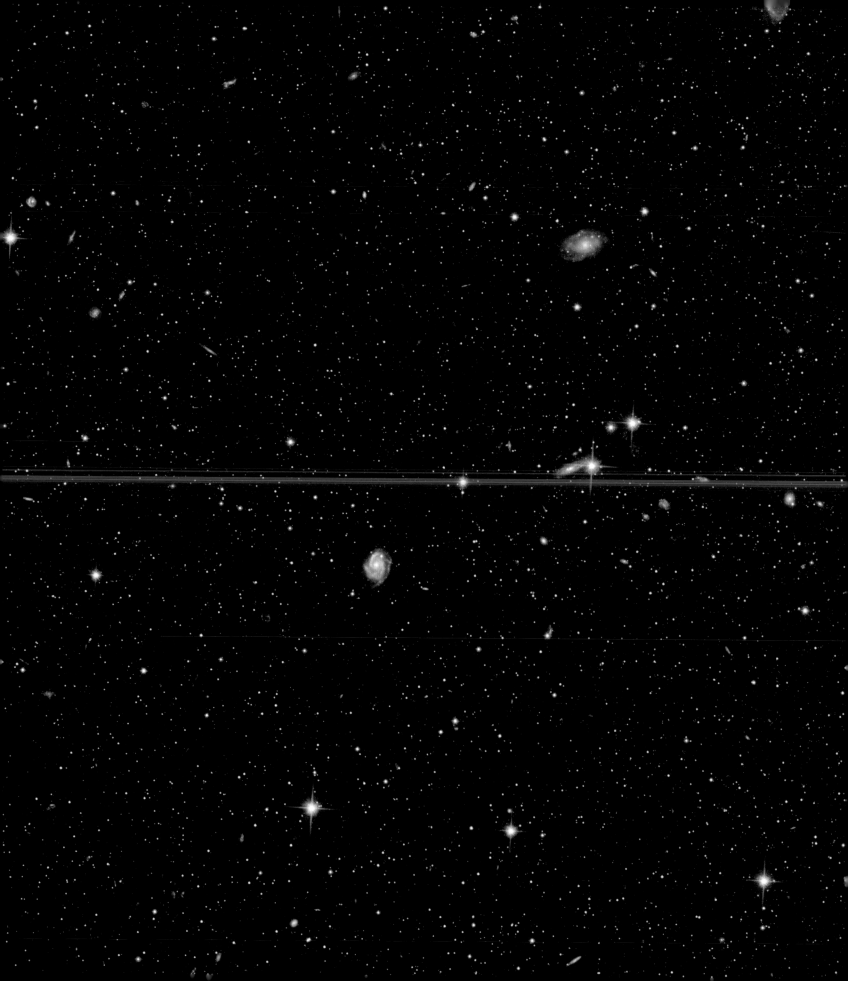

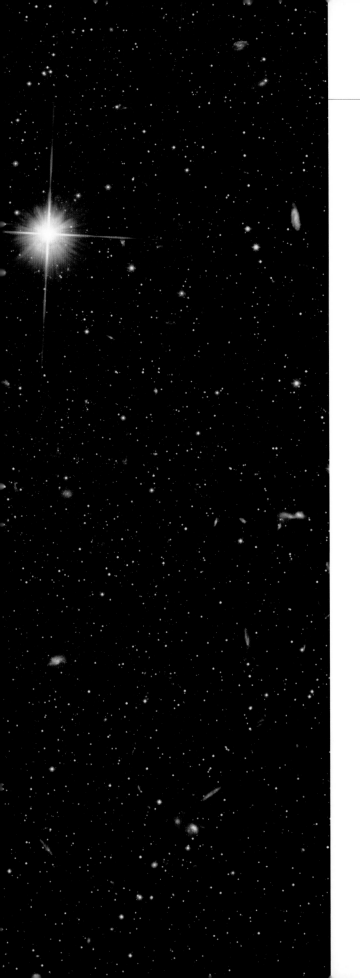

Probing the mysteries of massive galaxy halos: A deep view of the spherical halo of stars that surrounds the Andromeda galaxy, 2.5 million light years from Earth. Aside from revealing 300,000 new halo stars, the sharp Hubble Space Telescope image also shows up numerous background galaxies billion of light years away.

Enigmas and exotica

The story of astronomy is never ending. While giant telescopes, detailed calculations, and sophisticated computer models provide great advances in our understanding, they also raise exciting new puzzles. Space is rich in very strange and poorly understood phenomena which make the universe so wondrous. Some of the most challenging current conundrums, associated with some truly exotic objects, are collected in this chapter. We begin with the startling realization that a substantial proportion of the universe consists of elusive "dark matter" that is actually hidden from our telescopes. The race is on to identify this component, by examining detailed images of the distant universe and by developing state-of-the-art detectors buried in deep caves on Earth. The universe is also rocked by remarkably powerful bursts of gamma rays from enigmatic sources embedded in distant galaxies. Each of these awesome flashes can release more energy in a few seconds than our entire Galaxy emits in a decade. Scientists are also striving to construct a blueprint of the universe from the first three-dimensional maps of the largest structures known. Extensive surveys of millions of galaxies have revealed a "spongy" universe, characterized by vast sheets of galaxy clusters that stretch over millions of light years and are surrounded by even larger voids or holes. Finally, one of the most unexpected scientific discoveries of recent years is that the expansion of the universe is speeding up. This cosmic acceleration may be driven by a mysterious "dark energy" that counteracts the force of gravity.

This majestic spiral galaxy lies 100 million light years away in the constellation of Leo. Measurements of the manner in which spiral galaxies rotate about their centre provide valuable information relating to the amount of dark matter contained within the vast halos which surround spiral galaxies.

Hunting dark matter

Utilizing various independent experiments and methods, scientists have reached the astounding conclusion that more than 90 per cent of the universe exists in a form that is hidden from us. An important component of this enigmatic universe has been termed "dark matter". From an observer's viewpoint, dark matter is a mysterious constituent of space that cannot be seen directly, as it either does not radiate in any region of the electromagnetic spectrum or is too dim to detect with current technology. We know that dark matter exists because it has a gravitational effect on stars and galaxies; it is an invisible source of gravity which is considerably more abundant than the luminous matter we see around us.

Stellar tombstones such as collapsed white dwarfs, neutron stars, and black holes represent possible constituents of dark matter, on the grounds that when these objects are totally isolated in space they are extremely faint and undetectable. Another substantial component of dark matter may also take the form of highly exotic particles, very much smaller than an atom and weighing 100,000 times less than an electron. Although expected to exist in great numbers, these tiny particles rarely interact with ordinary matter and in some cases remain hypothetical, not yet detected in laboratories. We look here at how scientists are gathering evidence for the existence of significant amounts of dark matter, and their strides to discover its forms and impact. Our understanding of dark matter has a direct bearing on our knowledge of the evolution and fate of the universe.

Spinning galaxies

Magnificent spiral-shaped galaxies such as our Milky Way Galaxy have discs of stars, gas, and dust which rotate around their centre. Our Sun, for example, completes a single lap of the Galaxy once every 200 million years. Spiral galaxies do not spin like a giant solid plate, but move at varying speeds at different distances along the disc. By carefully measuring the speed of rotation of a galaxy at different distances from its centre, astronomers can learn a great deal about the nature of the mass of the galaxy itself.

It turns out there is something very strange about the speeds at which spiral galaxies rotate. The stars in a galaxy should be orbiting the centre in a similar manner to the way planets orbit the Sun in our solar system. That is, the further away a star is from the centre, the slower it should move (in the way that Neptune takes much longer to complete a lap around the Sun than does Mercury). Remarkably, measurements of the rotation of galaxies show that they do not slow down in the outer galactic regions as they may be expected to. In the Milky Way, for instance, stars orbiting at about 50,000 light years from the centre would travel at a speed of 540,000km per hour (335,000 miles

Looking toward the core of a globular cluster called 47 Tucanae, which contains groups of 12-billion-year-old stars and resides in the halo of our Galaxy. The light from a few of the numerous stars in 47 Tucanae may be briefly amplified by the passage of Jupiter-like brown dwarf stars in the foreground to the cluster. These gravitational lensing events show that brown dwarfs contribute to dark matter.

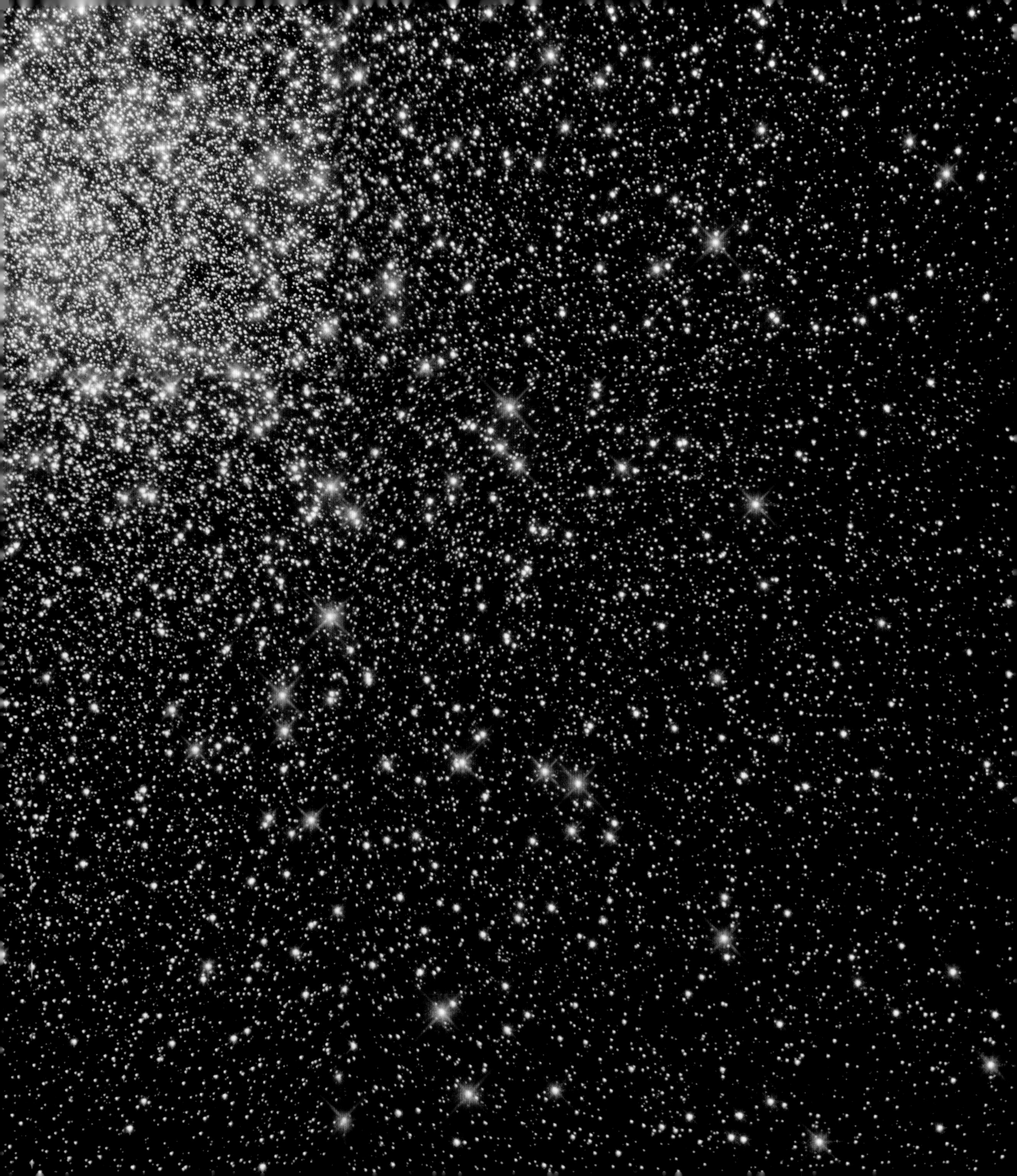

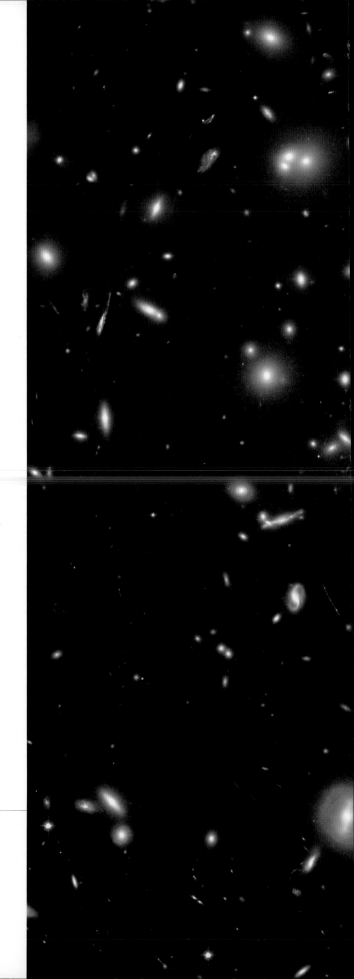

per hour) if they behaved like the planets in our solar system. In fact, measurements indicate that stars at this distance are travelling much faster, at 990,000km per hour (615,000 miles per hour). The problem is that when you total up all the visible matter due to dust, stars, and gas, there is not enough mass to hold these outermost stars in our Galaxy at such high orbital speeds. Moving this fast they should simply fly away and tear the Galaxy apart.

As spiral galaxies are not ripping apart everywhere, the explanation has to be that these outermost stars are held by the gravity of additional unseen material: galaxies must be surrounded by vast amounts of "dark matter" that cannot be observed by radio, infrared, optical, and X-ray telescopes. Measurements of our Galaxy suggest that only ten per cent of the total mass is in the form of stars, gas, and dust; an astonishing 90 per cent is hidden, dark matter, which is spread throughout a vast halo that surrounds all the galaxies in the universe.

As mentioned above, there are a number of possibilities for what the dark matter in galaxies might be. Astronomers have recently discovered evidence of "failed stars" called brown dwarfs. These are stars that have a mass that is less than ten per cent of the mass of the Sun, meaning that they are not substantial enough to start nuclear reactions in their cores, and are thus very dim and hard to see. Although there could be large numbers of brown dwarfs floating in the halos of spiral galaxies, it is very unlikely that there are enough of them to fully explain the peculiar rotation of galaxies.

Mass missing from clusters of galaxies

As we have discovered, galaxies are not randomly scattered across the universe, but are instead bound by gravity into large groups called "clusters". A cluster can have hundreds or even thousands of individual member galaxies. The Coma cluster is a typical example, lying about 300 million light years from Earth. It is loaded with more than 10,000 galaxies, including ellipticals,

The subtle effects of gravitational lensing are on view in this remarkable Hubble Telescope image of a cluster of galaxies known as Abell 1689, which is located more than two billion light years away. Hundreds of even more distant background galaxies are seen as faint, smeared and, distorted objects due to the gravitational bending of light by Abell 1689, which acts like a two-million-light-year-wide lens in space!

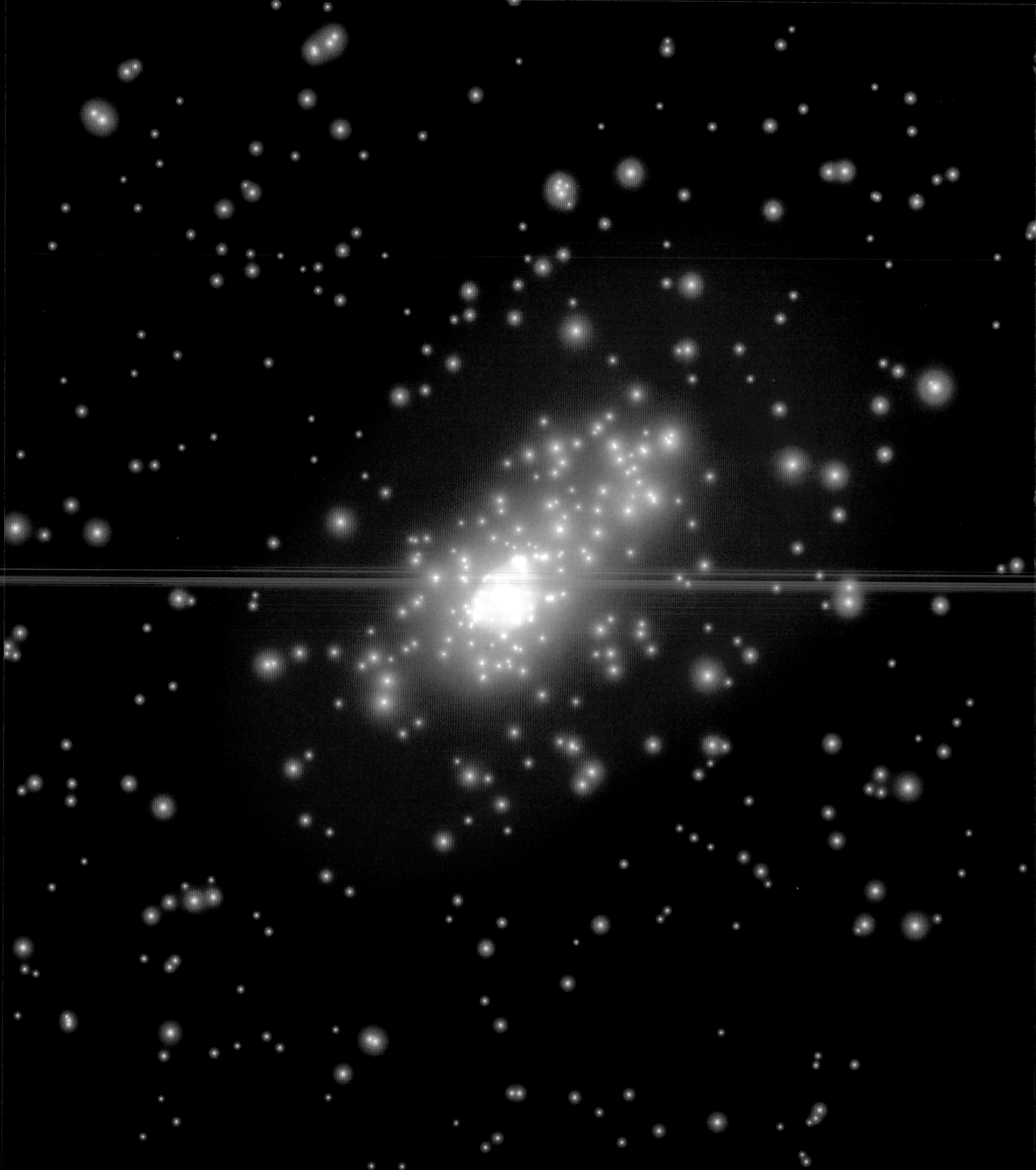

spirals, and dwarf types, and is experiencing lots of collisions and mergers. The manner in which individual galaxies in a cluster gravitationally tug on each other and move around depends on the mass of the galaxies and that of the cluster as a whole. Diligent studies of such galactic behaviour within a cluster make it possible to "weigh" the cluster as a whole. Astronomers can also use the technique of gravitational lensing to determine the presence and mass of dark matter as it, too, has a gravity and could act to bend light and form gravitational lenses. This is precisely what is happening in a galaxy cluster called Abell 1689, located more than two billion light years from Earth. Images taken by the Hubble Space Telescope reveal how the gravity of numerous galaxies, plus unseen dark matter, act as a powerful gravitational lens more than two million light years across. This object bends and amplifies the light from galaxies that are more than ten billion light years beyond it. These infant galaxies show up as arcs and smears of light superimposed on the image of Abell 1689. As the nature and quantity of these arcs is related to the manner in which mass is spread through the gravitational lens itself, they tell us how much dark matter must be present. The lensing in Abell 1689 in effect allows astronomers to map out how the dark matter is spread in the cluster.

The cosmic gravitational lens created by this cluster of galaxies 4.5 billion light years away has provided the basis for this unique map of luminous galaxies (in red) and the predicted spread of dark matter (in blue). The dark matter acts to hold the cluster together.

Clusters of galaxies also host extremely hot gas that radiates copious amounts of high-energy radiation. X-ray images of this gas have recently revealed that the hot medium is being held in place by dark matter, which prevents it from dispersing across the cluster and allowing it to cool more rapidly. Galaxy clusters are proving to be outstanding laboratories for understanding the properties of dark matter.

Ghostly particles from the Sun

Although the presence of dim and unseen objects such as brown dwarfs and black holes certainly accounts for some of the missing mass in galaxies and galaxy clusters, it cannot account for all the dark matter. Scientists agree that a substantial contribution to the dark matter component of the universe must also come from strange and elusive particles that are smaller than atoms and can pass right through ordinary matter. Sometimes called "weakly interacting massive particles" (WIMPS), they are thought to have a tiny mass, much less than that of an atom. The theory is that incredibly large numbers of WIMPS, and other strange particles, collectively make up a bulk of the missing matter in the universe. The search is on therefore to discover candidates for these ghostly particles.

One of the most recent advances has concerned particles called "neutrinos". Travelling close to the speed of light, these are similar to electrons, but carry no electrical charge. Neutrinos are very abundant – billions of them pass through our bodies every minute, but only one or two of them will actually interact with matter in our bodies over our lifetime. Scientists have been carrying out intriguing experiments to discover whether neutrinos have a mass at all; after all, if they are mass-less (like photons of light), then neutrinos would not make any contribution to the dark matter issue.

Our Sun is a vigorous nearby source of neutrinos, where they are produced during nuclear fusion reactions taking place in its core. They readily stream out because of their very limited interactions with the gas within the Sun. It is estimated that more than 10,000 billion neutrinos pass through each square metre of the Earth every second! The trick is actually detecting some of them, then measuring their minuscule mass.

Some success has been achieved by scientists working at a specially built neutrino detector called the Sudbury Neutrino Observatory (SNO), in Ontario, Canada. The detector itself is a 30m (33-yard) high tank filled with a thousand tons of very pure (heavy) water, and is located deep in a disused mine. The SNO's instruments are tuned to pick out rare interactions caused by neutrinos arriving from the Sun, which are detected as special flashes of light. Neutrinos actually occur in three different types and the latest results from SNO have shown that those arriving from the Sun are actually mutating between the different types, changing their character en route. This mutation from one type of neutrino to another is a complex process and is only possible if the particles are not mass-less.

The conclusion from state-of-the-art experiments such as SNO is that neutrinos do have a tiny mass, but it still remains difficult to determine what the precise value is. Even if the mass of a neutrino were just one 10,000th of the mass of an electron, this would still be significant, as there is such a great abundance of them in space. Future experiments are aiming to tightly constrain neutrino properties. Depending on exactly how massive they are found to be, they could account for a sizeable fraction of the dark matter in the universe.

Dark matter detectors

The search for other types of dark matter particles such as WIMPS is gaining pace, with the construction of new detectors designed to specifically pick up some of the most exotic particles in nature. One of the strangest laboratories has been built 1100m (1200 yards) underground in the Boulby mine in Yorkshire, England. The rock surrounding this deep mine shields the sensitive detectors from the radiation and normal particles, including cosmic rays, which continually bathe the Earth. The shield is no match for the WIMPS, however, which easily penetrate through it.

This sophisticated detector is based on liquid xenon, a rare chemical element that makes a suitable target for dark matter particles. The xenon has to be kept as pure as possible and at a precise temperature. Once fully calibrated and tested, remarkable experiments such as the one at Boulby, will cast new light on dark matter.

Viewed from the bottom, this is the vessel of the Sudbury Neutrino Observatory. Filled with 1000 tonnes of highly purified (heavy) water, the vessel is buried in a deep mine in Ontario, Canada, and is designed to detect the elusive neutrino particles streaming from the Sun.

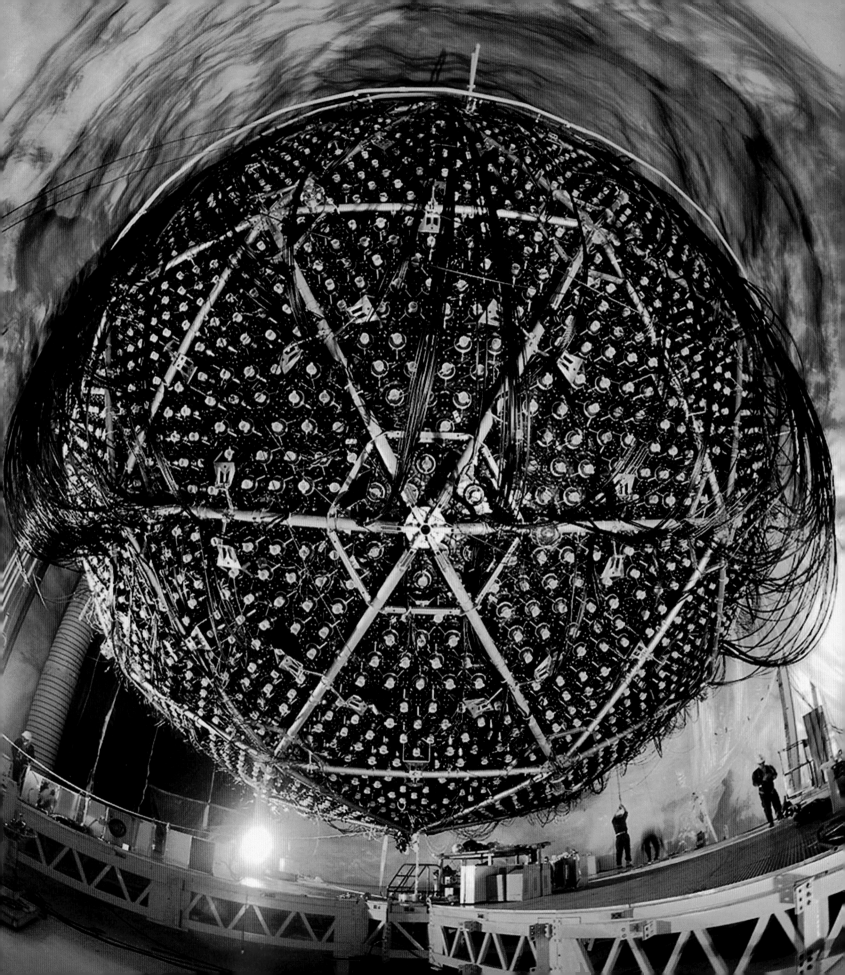

Gamma-ray bursts

One of the great controversies in modern astronomy concerns the origin of the incredibly powerful, short-lived explosions known as gamma-ray bursts. Originally discovered in 1973, these bursts last from just a few seconds to several minutes, and their perplexing sources are the most luminous emitters of energy in the universe. A single gamma-ray flash can, for a few seconds, outshine the rest of the universe. Only over the past few years have astronomers been able to deepen their understanding of this awesome phenomenon, thanks to the development of pioneering new telescopes based in satellites orbiting Earth.

Gamma rays are electromagnetic radiation of the highest energy, exceeding the energy of ultraviolet light and X-rays. Only the hottest sources of charged gas or plasma can produce gamma rays, which also makes these sites the most energetic and violent places in the universe. Despite having a greater penetrating power than X-rays, gamma rays are not able to get through the shield provided by the Earth's atmosphere. While this is great news in protecting us from harmful radiation, it also means that space-based observatories are required to study them. NASA's

An artist's impression of two neutron stars locked in an ever spiralling orbit around each other. It is thought that the ultimate violent collision between pairs of neutron stars may be the source of energy for some of the awesome gamma-ray bursts detected in the universe.

Compton Gamma-Ray Observatory was a complex satellite telescope which operated from 1991 until June 2000. Fitted with very sensitive instruments, during its lifetime the observatory detected and measured more than 2700 gamma-ray bursts and provided the first map of their locations in the sky. Unlike stars and gas which cluster in the disc and centre of our Galaxy, the Compton map as was shown by the mysterious gamma-ray flashes, were distributed all over the sky at unpredictable locations. When the gamma-ray bursts were first discovered more than 30 years ago, it was thought they probably arose from within our Galaxy. These new results from Compton demonstrated that the bursts come from all directions and have a cosmic origin which is in fact well beyond our Galaxy.

Pinpointing the explosions

In spite of the successes from the Compton Observatory, a key problem in understanding the source of gamma-ray bursts is that they are very difficult to accurately locate in the sky. A precise location is required in order to follow up the gamma-ray detection with normal, visible-light observations, in order to help identify what type of object resides there. The problem is made more difficult by the very short duration of the bursts, which means there is very little time from the initial detection of a gamma-ray event using a satellite and the triggering of optical telescopes like Hubble to take a deep image.

A significant breakthrough was achieved recently by the Italian- and Dutch-built satellite called BeppoSAX. Launched in 1996, it has the ability to identify gamma-ray bursts and locate

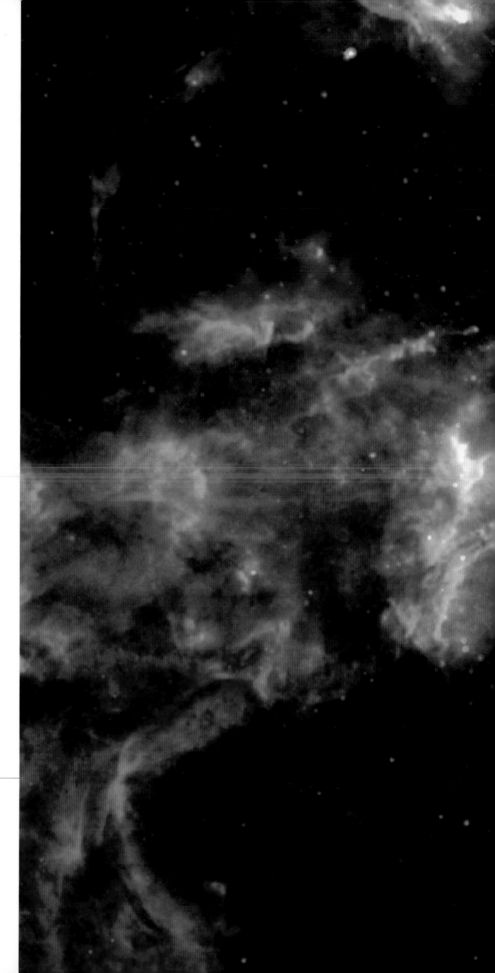

them very precisely in the sky, meaning that the flashes could be immediately chased by powerful ground-based telescopes. On 28 February 1997 BeppoSAX detected a gamma-ray burst (labelled GRB970228) toward the constellation of Orion. Within hours a X-ray image and several optical ones were obtained which revealed a dramatic dimming of the original source. The aftermath of the explosion revealed that a faint distant galaxy had hosted the burst. The precise alerts of gamma-ray flashes provided by BeppoSAX led to several follow-up discoveries of similar after glows, which could be seen for periods lasting from several days to several weeks in X-rays and visible light. The growing number of observations by the 10m aperture Keck telescope in Hawaii, and the Hubble Space Telescope, have now confirmed that the bursts are located in distant galaxies, some of which are more than ten billion light years away.

The fact that gamma-ray bursts are so distant and yet still appear bright in the sky means they must release astounding amounts of energy. Though each flash lasts barely a few seconds, the energy released is hundreds of times greater than that yielded in supernova explosions. Gamma-ray bursts clearly represent an extreme phenomena. The puzzle lies in understanding what kind of process is acting in their host galaxies.

Toward an understanding

The major challenge is to explain the circumstances which lead to these huge emissions of energy. The anatomy of a burst, derived from BeppoSAX observations and optical images of the after glows, consists of a powerful explosion that expands at near the speed of light, with the matter in the outburst cooling as it goes. What remains enigmatic is the precise origin of the energy that powers this violent outburst. A couple of currently mooted possibilities are considered here.

The most remarkable potential source for a gamma-ray burst is something called a "hypernova". This is essentially a scaled-up

The inner regions of a prolific star-forming region called the 30 Doradus nebula, viewed by the Hubble Telescope. Residing 170,000 light years from Earth, the sparkling highlight is the spectacular cluster of luminous stars seen to the upper right of the image. Known as R136, this cluster contains several dozen of the most massive stars known, some of which are 100 times the mass of the Sun and ten times as hot. The death of these phenomenal stars may be marked by powerful gamma-ray bursts.

version of a supernova explosion, which we met earlier as marking the catastrophic death of very massive, short-lived stars. Hypernovae are associated with the most active star forming regions in galaxies, and are triggered by the demise under gravity of the most massive stars in the universe. It's estimated that they release more than ten times the energy of a normal supernova explosion; the gamma rays are thought to arise from the impact between a fireball of matter moving at near the speed of light and a surrounding barrier of denser gas. Some observational support for this scenario has been provided by X-ray images of the giant Pinwheel galaxy (also known as M101) lying 25 million light years from Earth. Astronomers believe they have located the remnants of two hypernovae explosions in this galaxy. They stand out because of their intense X-ray emissions, and appear as vast shells of expanding gas. At more than 850 light years across, one of the remnants in the Pinwheel galaxy is the largest expanding shell ever discovered. The other shell is still bloating out at more than 360,000km per hour (225,000 miles per hour). There remains a lot to be discovered and understood about hypernovae, but they could possibly be the most violent explosions since the Big Bang itself. It seems they are signatures of the most spectacular stellar deaths known, which makes them a likely source for gamma-ray bursts.

Another exotic possibility is that some of the flashes of gamma rays may be linked to neutron stars. Recall that a neutron star is created from the stellar core left over after a massive star explodes in a supernova. The core which becomes a neutron star is rapidly spinning, and so dense that it is only about the size of a city! Neutron stars also have very strong magnetic fields and an interior composed of matter which has strange electrical and frictional properties. Amazingly the tiny stars also have a very thin crust, and it is thought that quakes, analogous to earthquakes and driven by the intense magnetic field, buckle the crust. This results in the propulsion of jets of high-speed charged particles, which then produce gamma rays.

Scientists believe that gamma-ray bursts could also arise from rare collisions between two neutron stars. The theory is that some neutron stars are part of a binary star system, in which a pair of collapsed stars orbit each other at great speeds, sometimes completing hundreds of orbits in a second. The close orbits shrink further and the two high-density stars eventually spiral into each other. Through mechanisms that are still not fully understood, the tremendous energy from the collision of the two stars is translated into a gamma-ray burst.

A new generation of gamma-ray telescopes will offer the promise of delivering a deeper understanidng of these awesome explosions.

The massive supergiant star Eta Carina is destined for destruction in a few million years from now. Born with a mass of more than 100 Suns, this remarkable object may be a candidate for a hypernova explosion, resulting in a powerful gamma-ray burst. The optical (left of centre) image reveals spectacular bubbles of expanding gas that signal the unstable state of Eta Carina. The inner regions of this turbulent outflow are also imaged here in the X-rays by the Chandra satellite. Material travelling at supersonic speeds is creating awesome shock waves that heat the surrounding gas to millions of degrees.

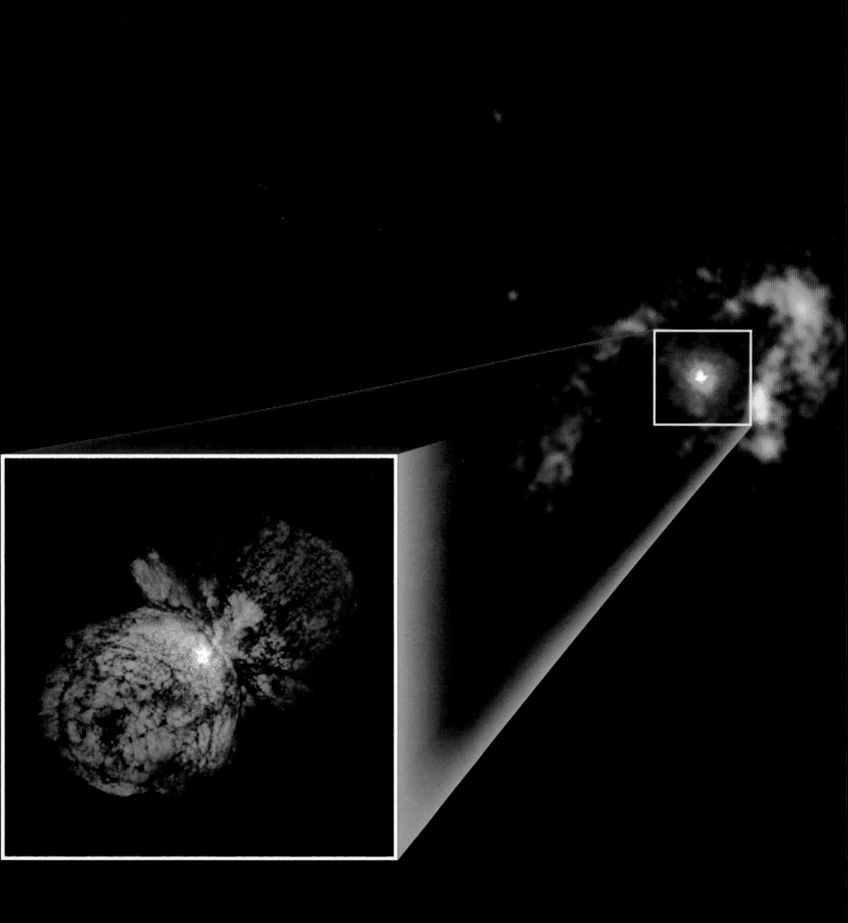

The grand design

A rapidly developing field of study in astronomy is the quest to understand how galaxies are distributed throughout the universe. The latest generation of large telescopes is being used to map out the spread of vast populations of galaxies, stretching over hundreds of millions of light years. The goal is to understand the grand design of the universe itself and unravel its earliest evolution. We chart in this section the steps that astronomers are taking toward mapping the three-dimensional distribution of galaxies.

We have already seen that galaxies rarely occur in isolation, but are most often found in clusters or groups that are bound by their mutual gravitational pull. Such clusters can themselves be grouped into a hierarchy based on size. Our Milky Way Galaxy is a member of a small cluster of about 40 galaxies called the Local Group, which spans a few million light years, although richer clusters can contain hundreds of galaxies. The largest near to us is the Virgo Cluster, named after the constellation in which we see it. This hosts more than a thousand galaxies and stretches beyond ten million light years. Remarkably, there are even heftier gatherings of galaxies in space, the cluster known as MS1054-0321 being one of the most extreme. Located almost eight billion light years from Earth, and dating to a time when the universe was half its present age, it contains thousands of galaxies, each swarming with billions of stars. Besides vast amounts of dark matter, this cluster also contains some of the hottest gas ever studied. X-ray observations have provided images of vast clouds of intergalactic gas which has been heated to 150 million °C (300

million °F). It is of great importance to cosmology that such a substantial cluster of galaxies was already in place at such an early phase in the evolution of the universe.

The largest hierarchal size is defined by super-clusters, which represent the groupings of hundreds of clusters, spanning a distance of a hundred million light years. The Local Group of galaxies and the Virgo Cluster, for example, are part of a vast local super-cluster. Super-clusters form complex networks that permeate throughout space and characterize the large-scale structure of the universe.

Surveying the universe

In order to learn more about the universe on the largest scales, astronomers have initiated several impressive projects designed to survey vast numbers of individual galaxies over distances extending 500 million light years. One example is the 2dF Galaxy Redshift Survey, which is a large international collaboration between scientists from several countries. In order to construct a three-dimensional map of the universe, the 2dF project has secured measurements of the positions and distances of 250,000 galaxies.

Known as the Minuet, the members in this group of galaxies are named after the graceful way they orbit around each other, over periods of hundreds of millions of years. In a similar manner, our own Milky Way Galaxy is part of a "Local Group" of about 40 gravitationally locked galaxies.

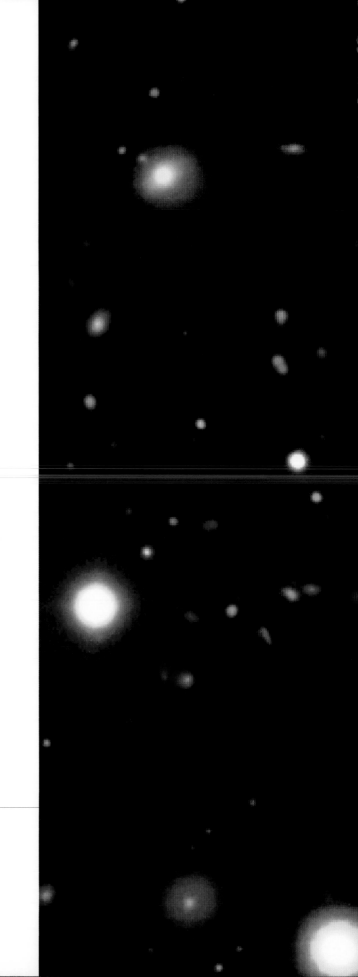

The results from the 2dF study and other similar projects are astonishing. They reveal a "spongy" or "bubbly" structure to the universe, with great clusters of galaxies arranged in thin sheets or long filaments. Dispersed among the sheets and filaments of super-clusters are vast voids or holes, which lack detectable luminous matter and are typically hundreds of millions of light years across; the largest one known, called the Bootes void, has a diameter of some 400 million light years. These voids are a dominant feature of the galaxy maps and account for almost 90 per cent of space. One of the great challenges of cosmology is to understand the origin of these remarkable structures.

Extensive galaxy surveys have revealed some truly exotic objects. Among the largest known single structures in the universe is a sheet of galaxies referred to as the Great Wall. Located about 250 million light years from Earth, the Great Wall is 500 million light years long and 200 million light years wide, but with a thickness of only 20 million light years. The bulk of it is composed of a super-cluster of galaxies that contains ten million billion times the mass of the Sun. Another strange beast, dubbed the Great Attractor, is a mysterious object that is apparently pulling in millions of galaxies in a patch of the universe that includes the Local Group (including our Milky Way Galaxy) and other neighbouring clusters such as the Virgo Cluster. The Attractor itself is some 250 million light years away in the direction of the constellation of Centaurus. Galaxies are being dragged toward it at 360,000km per hour (225,000 miles per hour). This velocity suggests that this great unseen mass may contain ten times more matter than all the visible matter in that direction.

These incredible structures and their movements cannot be explained as part of the overall expansion of the universe. The vast sheets of galaxies and the adjacent voids are far too large to have been created by the straightforward action of gravity. Despite its 14-billion-year-age, the universe is simply not old enough for gravity alone to have gathered ordinary matter and distributed it in such gigantic filamentary and bubbly patterns. A blueprint must have been in place from the earliest epochs of the universe that forced matter into this grand design.

Peering a distance of more than eight billion light years, this remarkable Chandra satellite X-ray image reveals the most massive cluster of galaxies ever observed at such an early stage in the history of the universe. The purple shading highlights 70 million°C (160 million°F) hot gas, confined within the cluster's total mass of 200 trillion times that of the Sun!

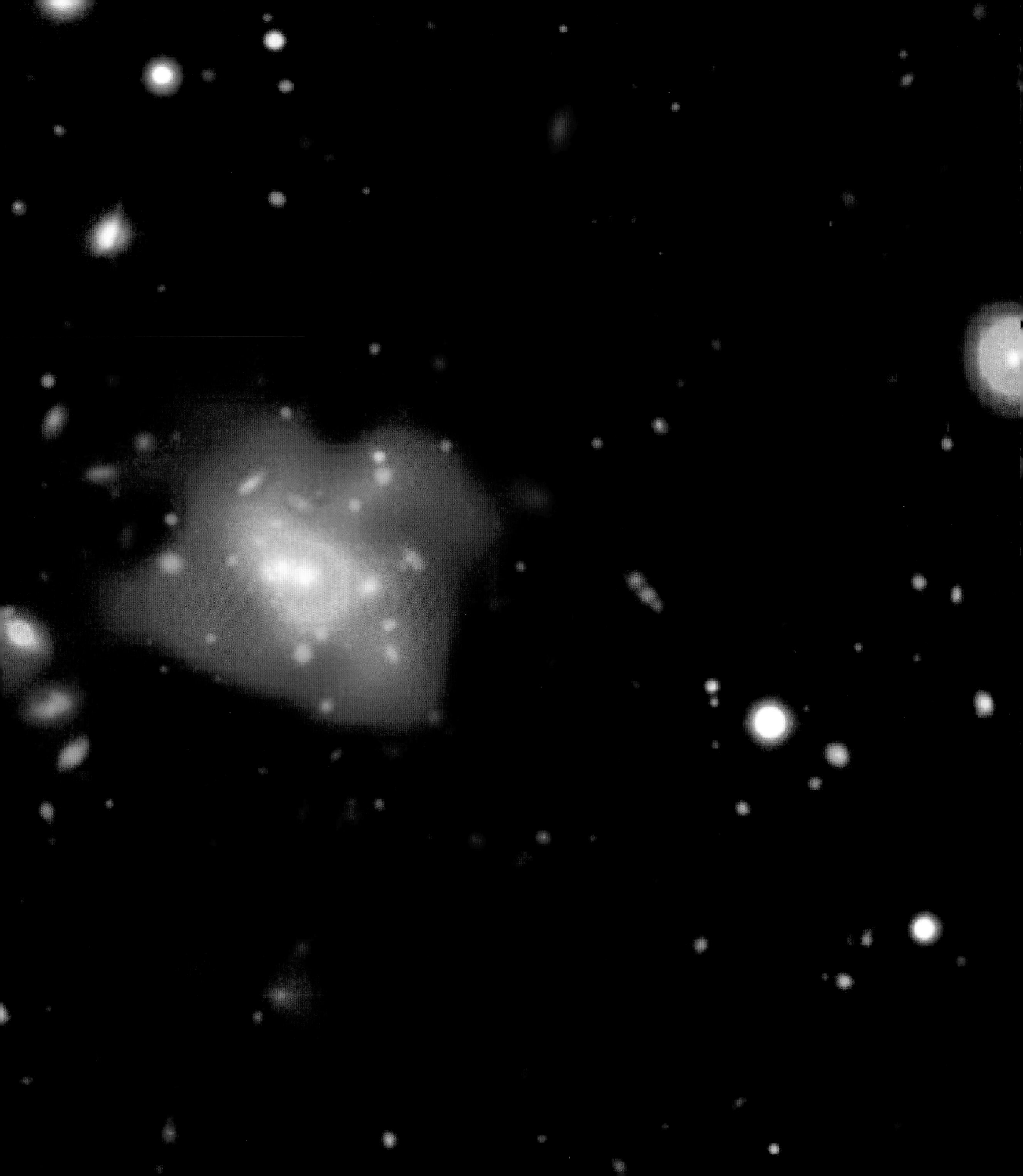

Growing from seeds

In order to explain the large-scale structure of the universe, astronomers have been experimenting with supercomputers, to simulate how structures in it could be affected by mixtures of different kinds of matter that may have been in place at the earliest times after the Big Bang. Two basic types of dark matter in particular are used, both relating to properties of subatomic particles. First is "hot dark matter", which is made of particles that weigh much less than an electron, such as the neutrinos described earlier in this chapter. The effect of including too much hot dark matter in the computer models is to create the large-scale features such as super-clusters and voids, but not enough of the smaller structures which are also observed. The alternative ingredient is "cold dark matter", composed of more massive particles which more easily produce small-scale structures, including individual galaxies. By adjusting the mixture of these two types of dark matter it is possible to simulate the growth of tiny concentrations of matter in the early universe. These fluctuations in the density of matter over small regions of space act like seeds that grow over billions of years with the expanding universe. It is encouraging that the calculated predictions of this evolution compare very well with the maps of actual large-scale structure provided by extensive galaxy surveys.

The currently favoured scenario is a universe dominated by cold dark matter in which the individual galaxies formed first. They then grouped into clusters and super-clusters, with most of the dark matter concentrating in the great voids. A lot of work still remains to be done, however, to deepen our understanding of the design of the universe; indeed, the particles proposed to make up cold dark matter have not yet been detected. We are nevertheless entering an era where substantial new observational surveys will provide three-dimensional positions for millions of galaxies. These projects are certain to present new paradigms, more bizarre objects, and alternative theories about the grand design on view.

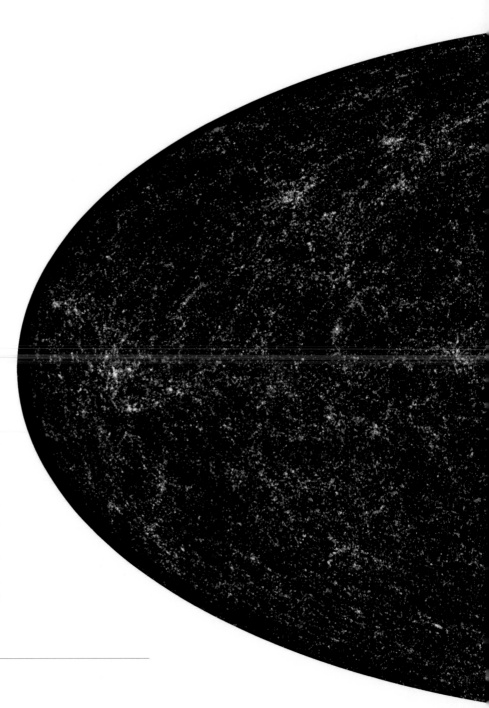

This panoramic view from the 2MASS satellite-based survey shows the spread of more than a million galaxies over all directions in the sky. The brightest and nearest galaxies are displayed in blue, and the faintest and most distant ones are in red. The colour scheme therefore provides an insight into the three-dimensional large-scale structure of the universe, which is defined by filamentary patterns (of galaxies) surrounded by enormous dark voids (deficient of any galaxies).

The universe is in overdrive

Perhaps one of the most astounding and mysterious discoveries recently made by astronomers relates to the manner in which the universe is expanding. By studying very distant exploding stars, scientists have found evidence that we live in a universe that is growing at a faster and faster rate. The latest findings are contrary to what might be expected from standard ideas of the Big Bang origin, and they raise the spectre of a mysterious form of energy called "dark energy".

During the 1920s the astronomer Edwin P. Hubble discovered that galaxies are moving away from us, and away from each other, with the implication that the universe as a whole is expanding. This fundamental observation provided direct support for the Big Bang model of the universe. Since that time two of the key questions that cosmologists have sought to answer are how long the expansion will continue and by what means the expansion will proceed. Intuition would suggest that following the violent event that first propelled the universe 14 billion years ago, the attractive force of gravity would eventually start to hold matter back, thus slowing the galaxies and the expansion of the universe itself. There are, after all, billions of galaxies exerting a gravitational pull on the space around them, and it is reasonable to expect or hypothesize that they will act to counter cosmic growth. Scientists have therefore been trying to measure the degree to which the expansion is slowing down. Contrary to their expectations, however, what they have in fact recently discovered regarding the universe's rate of expansion is completely unexpected and may have potentially far-reaching consequences.

Beacons and bulbs

In order to study how the expansion of the universe is changing it is necessary to sample light from the most distant galaxies. As this light takes billions of years to reach us, we are in fact observing galaxies as they appeared when the universe was much younger, just a fraction of its present age. If gravity was indeed slowing down cosmic expansion, then the distance between us and a very remote galaxy would be less than if the expansion had continued at a constant rate, and the galaxy would consequently appear brighter because it would be closer to us. As an analogy, imagine two identical cars speeding away from a set of traffic lights for 20 seconds. Both cars travel at the same speed for the first ten seconds, but then one of them slows down over the next ten seconds, while the other continues at the constant original speed. At the end of the 20 seconds, the car that slowed down will, of course, be closer to the starting position, and as a result the traffic lights behind will appear brighter than to the car that stayed at the same speed.

The difficulty is in finding objects billions of light years away the distances of which can be so accurately determined that it becomes possible to predict how bright they should appear

A Type Ia supernova explosion is seen at the lower left in this Hubble Space Telescope image of the galaxy NGC4526. The galaxy (and supernova) is located 108 million light years from Earth. For a brief period the supernova was comparable in brightness to the entire galaxy.

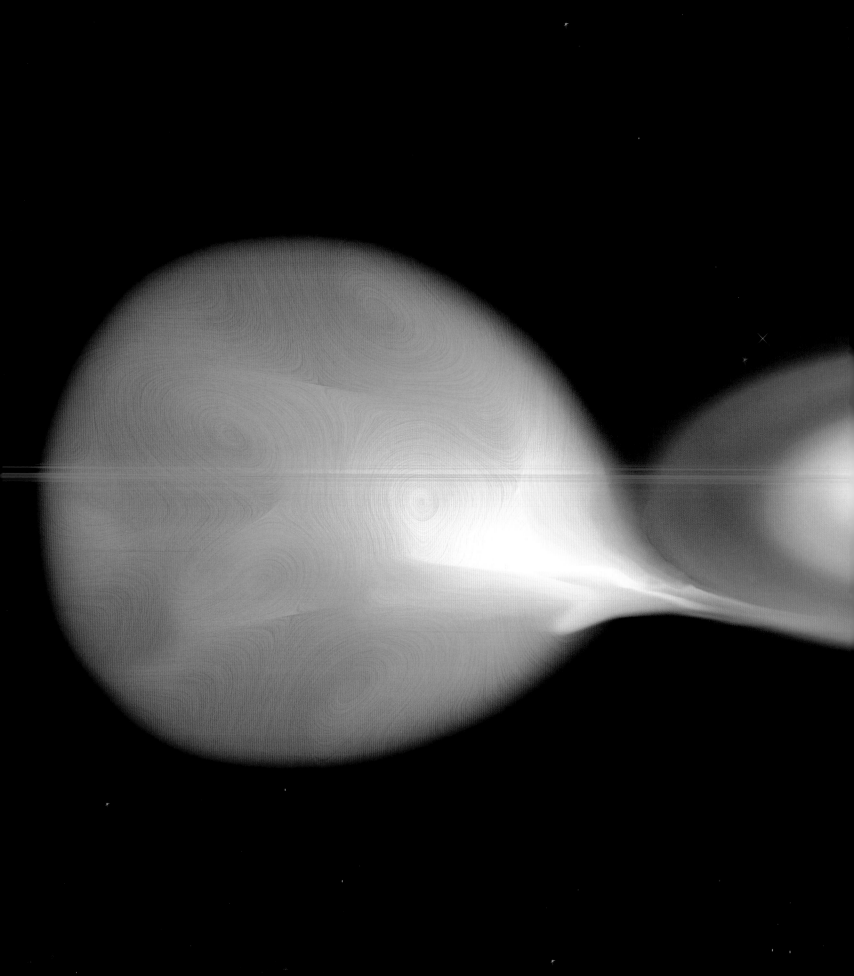

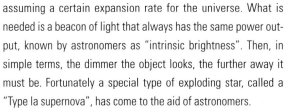

assuming a certain expansion rate for the universe. What is needed is a beacon of light that always has the same power output, known by astronomers as "intrinsic brightness". Then, in simple terms, the dimmer the object looks, the further away it must be. Fortunately a special type of exploding star, called a "Type Ia supernova", has come to the aid of astronomers.

The supernovae explosions we have encountered so far, called "Type II supernovae", signal the destruction of the most massive stars once they have exhausted all possibilities of nuclear fusion energy. There is, however, another manner in which a star can explode. In this alternative case a dying white dwarf star pulls huge amounts of gas onto itself from a very nearby companion star, like a red giant. The result is that the white dwarf becomes too massive to support itself against gravity and undergoes an enormous thermonuclear explosion. The brightness of the star increases by a billion times for a brief period, then fades gradually over a period of a few weeks and months. This is known as a Type Ia supernova, and at its peak the destructive act can appear as bright as its entire host galaxy. Type Ia supernovae are thus superb beacons over great distances in the universe. The crucial advantage is that all Type Ia supernovae are thought to be very similar. They are like incredibly powerful identical light bulbs of the same wattage. One problem is that these explosions are rare, and dedicated monitoring programs are required to find them and study changes in their light.

The universe is speeding up

For the past few years teams of astronomers have been using giant telescopes to locate Type Ia supernovae in very distant galaxies. The most distant ones exploded more than ten billion light years away.

The expectation was that the universe is today slowing down, due to the action of gravity on all the matter contained within it. The startling surprise from supernova hunting projects was that the Type Ia supernovae were further away than was expected. Instead of appearing brighter because the universe is decelerat-

An artist's view of a system of binary stars that can lead to a Type Ia supernova explosion. The compact white dwarf on the right is pulling in material through a swirling disc, from the normal giant star on the left. This accretion of matter onto the white dwarf can lead to a phenomenal thermonuclear explosion that can be detected in very distant galaxies.

ing, the ancient supernovae appear almost 20 per cent dimmer than expected. The implication is that the growth of the universe has sped up over the past few billions of years. The universe is more revved up today and its expansion is accelerating, thereby increasing – instead of decreasing – the distance between the Earth and the galaxies hosting the stellar explosions. Let us return to our analogy of the two cars speeding away from the traffic lights. Now imagine that instead of slowing down after ten seconds, one of the cars accelerates for the final ten seconds. This means it will be much further away after the full 20 seconds compared to the original cars, and of course looking back the traffic lights will now appear fainter. The observations of supernovae also suggest that a transition has occurred between the universe initially slowing down after the Big Bang and subsequently speeding up. More data are needed to date this transition accurately, but current estimates place it at a time when the universe was about a third of its present age.

A mysterious force?

The fact that Type Ia supernovae are much further away than originally expected raises the question of what force is pushing everything apart faster now than it did in the early universe. No one really knows the answer, and scientists have proposed a few different theories and ideas. One commonly debated possibility is that the growth of the universe is being accelerated by a mysterious form of energy called "dark energy", though there are almost no clues as to its identity, and it certainly does not emit any light. This unusual form of energy must permeate throughout the uni-

verse and act to counter the effects of gravity. It is a repulsive force, like a cosmic anti-gravity that is capable of overcoming the attraction of matter over vast distance, therefore accelerating the expansion of the universe. Astronomers believe that almost 70 per cent of the universe is composed of dark energy, which means it certainly cannot be disregarded. Understanding the destiny of the universe thus requires an understanding of dark energy; the hunt is on to determine its nature and origin. One of the most critical questions to answer is whether dark energy is unchanging in time: if it stays the same then the universe will certainly continue to expand forever.

Unlocking these mysteries requires synergy between different fields of science, including cosmology and particle physics. It is unlikely that dark energy could be produced in a laboratory, at least not in the near future, so the most probing and direct clues as to its nature will undoubtedly come from deeper views of the universe, which reveal not just distant supernovae, but also the intricate temperature patterns in the cosmic microwave background radiation. It may even turn out that the presence of dark energy impacts on some of our fundamental constants, such as the strength of gravity, so that even they are variable in time.

The turbulent debris of a Type Ia supernova explosion first observed by the astronomer Tycho Brahe in 1572 is viewed in detail by the Chandra X-ray satellite. This example of stellar remains is a mere 7500 light years from Earth, which is very nearby compared to similar explosions detected 10,000 million light years away that provide evidence for an accelerating universe.

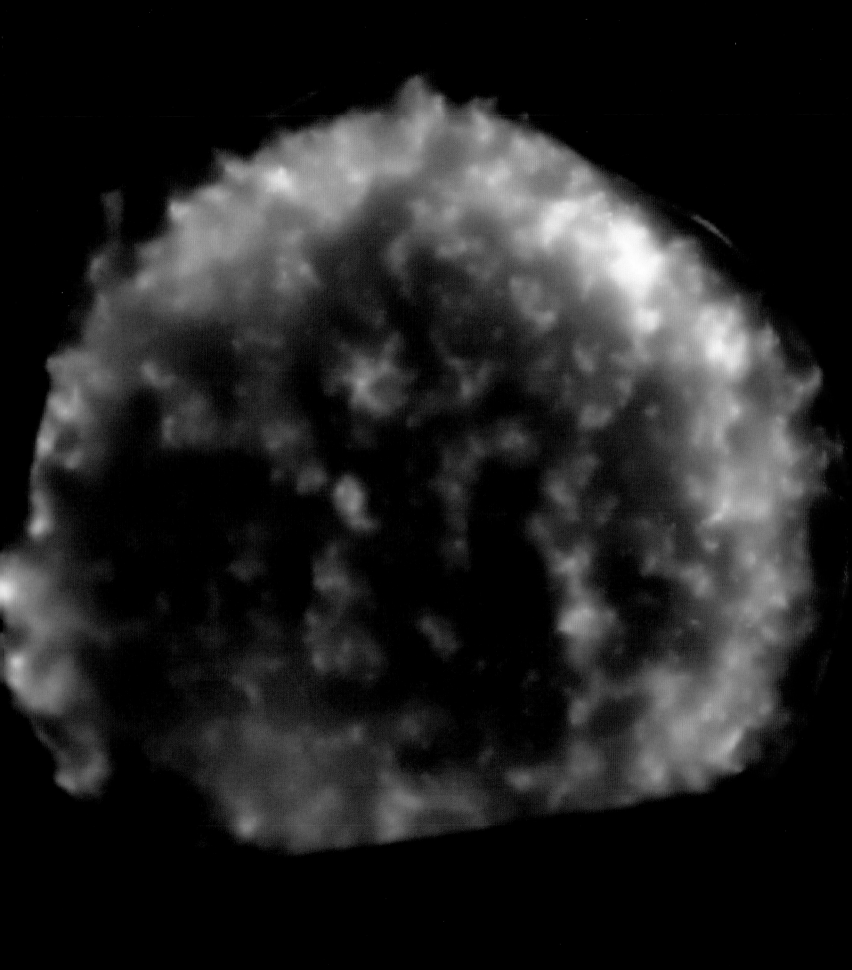

Looking beyond Hubble: An artist's impression of NASA's powerful successor to the Hubble Space Telescope. Named the James Webb Space Telescope, it could begin operations by 2011.

The view ahead

The spectacular discoveries revealed in this book have expanded our understanding of the universe, but there remain intriguing and fundamental questions that still need answers. The overarching quest is to develop our knowledge of how the universe evolved from the Big Bang to the formation of our own planet. The strive toward this goal will demand a detailed survey of the universe and its constituents, including the enigmatic dark matter and dark energy.

Astronomers will aim to peer deeper in to the era when the first stars and galaxies formed. They will look to unravel the life stories not only of stars but also of giant and terrestrial planets, leading to a hunt for the "golden prize" marked by unambiguous evidence for extraterrestrial life. The view ahead is of a universe that is a unique laboratory for probing the laws of nature. Progress in improving our knowledge is closely linked to our ability to detect light from the faintest, most distant objects in space. The great advances of the coming decades will demand exciting technological innovations to permit the construction of fantastic new telescopes and sensitive instrumentation. New observatories are now being designed with the capability to deliver enormous strides in our understanding, from the evolution of galaxies and stars to the fingerprints of life in the atmospheres of extra-solar planets. The grand ground-based telescopes and spacecraft missions of the next generation are introduced in this final chapter.

As depicted in this artist's rendition, a long-term base on Mars is one of the exciting goals currently proposed by NASA and ESA. Several studies have now been initiated to investigate possibilities for crewed missions to the Red Planet. The primary components of outposts there might include habitat modules to live in, exploration vehicles, power plants, and water pumping stations.

Mega-telescopes

Astronomy is largely an observational science, and its principal instrument is the telescope. Current-generation premier telescopes, operating over a wide range of the electromagnetic spectrum, provide impressively detailed and distant views of the cosmos. The telescope is essentially a "light bucket", its principal function being to collect as many photons of light as possible. This means the larger the telescope is in diameter, the better – giant telescopes gather more light, which allows astronomers to study bright objects in greater detail, as well as record the faintest and more distant targets. The current standard of telescopes sited on Earth is set by facilities such as the 10m (11-yard) diameter Keck telescope on the island of Mauna Kea in Hawaii, and the suite of 8m (9-yard) telescopes operated by the European Southern Observatory in Chile. Highly successful space-based observatories such as the Hubble Space Telescope have the crucial advantage of escaping the blurring effects of the Earth's dynamic atmosphere, as well as the ability to sample non-visible radiation that is largely absorbed before reaching the ground.

Plans are now in place for outstanding new projects that will deliver the next generation of telescopes, capable of greatly advancing our knowledge of space. We take a look in particular at an exciting successor to the Hubble Space Telescope, and a gargantuan 100m (109-yard) diameter ground-based telescope.

The Gemini North Telescope perched on the island peak of Mauna Kea, Hawaii. This is one of the current generation of 8m (9-yard) diameter telescopes. Some of the telescopes planned for the next decade will dwarf the Gemini instrument, providing vastly increased light-gathering capabilities.

Previous pages: The tremendous success of the Hubble Space Telescope has in a major part been due to the fact that its camera, spectrographs, and other components can be upgraded by visiting astronauts. This scene from March 2002 shows astronauts from NASA's *Columbia* shuttle installing a new power unit on the telescope, during a seven-hour space walk.

Hubble's successor

There is no doubt that the exquisite images from the Hubble Space Telescope have transformed our view of the universe, and it now rates as one of the most successful and advanced orbiting telescopes ever built. This prolific instrument is, however, only expected to continue operations until 2007 or 2008, having by then logged almost two decades of continuous service. Astronomers have therefore already begun designing its successor, which will be larger and even more powerful.

The new telescope has been named the James Webb Space Telescope (JWST), in honour of James E. Webb, the head of NASA from 1961 to 1968. The most important objective for JWST is to peer at light from the very first stars and galaxies created in the universe. It will target a critical period of time in the universe about which we currently know very little, referred to as the "dark ages of the universe". Spanning the period between a million and a few billion years after the Big Bang, it was during this time that the first stars and galaxies began to form. The reason the Hubble Telescope cannot already detect the distant objects from this time period is that light emitted by the youngest stars and galaxies is shifted in wavelength, from the visible (where Hubble primarily operates) to the infrared region. This shift in the colour of light (known as redshift) is caused by the expansion of the universe from the "dark ages" to the present time. The JWST will in effect have infrared eyes, which will also permit it to directly address several other important processes in space, such as the path by which galaxies evolve over billions of years and the manner in which planets are formed.

The telescope itself will be state-of-the-art and as of July 2003 is being developed in a partnership between scientists and engineers in the United States, Canada, and Europe. It will have a diameter of 6m (20ft), compared to the 2.4m (7.9ft) aperture of the Hubble Telescope. To operate successfully in the infrared region of light, heat from the telescope itself, which would radiate in the infrared, must be eliminated, and therefore its instruments and detectors will have to be cooled to -223°C (-370°F) throughout its projected mission lifetime of five to ten years. Unlike Hubble, the JWST will not benefit from regular servicing by astronauts ferried on the space shuttle, as it will be placed in a considerably more distant orbit 1.5 million km (940,000 miles) from Earth. Provided the funding remains in place, JWST's 5400kg (12,000lb) load will be lifted from Earth using a medium-sized rocket some time during 2011.

The most advanced technologies available are expected to provide JWST with remarkable imaging capabilities, which, when combined with the larger light-gathering ability and its dark, stable location in space, promise to deliver an awesome new facility. Astronomers eagerly await the marvels to be revealed.

The overwhelmingly large telescope

The most ambitious current plan is for the construction of a new ground-based telescope that will be 100m (109 yards) across. Known as the Overwhelmingly Large Telescope (nicknamed OWL), its light-gathering ability will be a dramatic hundred times greater than the premier telescopes currently in use today. This astounding instrument, proposed by the European Southern Observatory, stands to revolutionize astronomy.

The toughest challenges relate to the construction of the 100m (109-yard) primary mirror needed for OWL. Approaching the size of a football pitch, it cannot be cast from a single block, but will instead be comprised of more than 2000 hexagonal mirrors that are "stitched" together. Each of these segments will be no more than 2m (2 yards) across, with a complex system of sensors and pressure pads behind them. The sensors and pads will be adjusted by computer control to maintain the perfect shape for the overall 100m structure.

One of the major frustrations for all ground-based telescopes, no matter how large they are, is that the Earth's atmosphere is always in motion. Observing astronomical objects through this turbulence degrades the detail of the images on view, which is why a telescope placed in orbit has such a great advantage, as demonstrated by the Hubble Space Telescope. The OWL, however, aims to beat the distorting effects of the atmosphere by using radical technology called "adaptive optics". In essence, adaptive optics make it possible to eliminate the effects of atmospheric turbulence by using the light from a bright reference star, or a laser beam shone into the sky, to provide a guide as to how distortions affect the image. This information is rapidly delivered to a small additional mirror the shape of which can be altered to compensate for the blurring. Such an advanced adaptive optics system is key to the success of the OWL project.

The OWL is expected to give astronomers unprecedented views of the most distant objects in the universe and reveal fine details of stars and galaxies with a clarity 40 times greater than the Hubble Space Telescope. It will be able to detect the light from the most remote supernovae explosions to provide a distance yardstick over even greater spans of the universe. Closer by, it will enable scientists to study the surface regions of other stars in the manner that we currently study our own Sun. It is hoped that OWL will also provide new perspectives on the chemical composition of extra-solar planets, and contribute to the hunt for life-supporting environments beyond our solar system.

These golden goals will not come cheap, however. The European Southern Observatory estimates that OWL will cost around a billion euros (more than one billion US dollars), and a substantial international partnership is being gathered to try to meet this level of funding. Scientists and engineers are confident, however, that this giant telescope can be built, and there is anticipation that OWL may start delivering its first spectacular images by 2015.

Following pages: An illustration of the phenomenal 100m-diameter OWL telescope planned by the European Southern Observatory (ESO). The new telescope could be providing exquisite views of planets, stars, and galaxies by 2015.

Great innovations in space

While both the James Webb Space Telescope and the Overwhelmingly Large Telescope are clearly substantial new facilities for the next decade, they rely on concepts and technologies that are largely already available . There are, however, greater plans afoot that will prove even tougher challenges and demand the most sophisticated innovations.

Two specific missions discussed in this section have in common the notion of combining the signal from a flotilla of detectors spread over substantial distances in space. Known as "interferometry", this is a clever technique to boost the resolving power of the telescope, or its ability to discern the finest details. Where the detectors used in these strategic arrangement are telescopes, their coordinated effect is to deliver the power equivalent to what would be expected from one gigantic instrument, with a diameter equal to the distances between the individual "stations".

Hunting distant Earths

The more than 100 planets so far discovered around other stars are all believed to be Jupiter-like gaseous planets, lacking both a solid surface and liquid water. A hugely significant goal for the future is to find Earth-like planets capable of supporting life. The European Space Agency (ESA) is now planning an exciting mission named Darwin, after the British naturalist, with the primary objective of detecting and analysing such planets around nearby stars. The plan is to launch six telescopes, each 1.5m (5ft) across, on a version of ESA's 52m (57-yard) high Ariane 5 rocket, which is designed to eventually carry payloads of up to 12 tonnes. The telescopes will be flown to a distant orbit 1.5 million km (1 million miles) from Earth, where they will be arranged into a circular configuration, each separated by 50m (55 yards). A central hexagonal spacecraft will collect and combine the light from the six individual telescopes, thus providing significantly increased resolution and much more detailed images compared to other space-based telescopes. Laser beams will be directed between the six telescopes to keep the entire configuration in the correct alignment, to an accuracy of a thousandth of a millimetre. One advantage of Darwin is that it will operate in the infrared, rather than the optical (visible), spectrum. The benefit is that, while a Sun-like star outshines an Earth-like planet by a billion to one in visible light, this drops to a million to one in infrared, making the detection of planets slightly easier.

Darwin aims to survey a thousand of the nearest stars in the hunt for previously unseen small, rocky planets. The incredible sensitivity of this interferometric instrument will enable astronomers to capture the first images of Earth-like worlds as points of light. Having discovered their light, the composition of the planets' atmospheres can subsequently be examined using a spectrograph, which splits light into its constituent wavelengths.

The telescopes of the European Space Agency's Darwin mission are depicted here. The primary goal of this exciting mission, and others such as NASA's Terrestrial Planet Finder, will be to search and analyse the faint light from Earth-like planets orbiting around distant Sun-like stars.

This technique allows deeper analysis of the gases present in the atmospheres and makes it possible to determine whether life may exist. The presence of oxygen in a planet's atmosphere, for example, is a key diagnostic, as carbon-based life is the only known source for a continual input of this element into a terrestrial planetary atmosphere. Oxygen is the by-product of photosynthesis by bacteria in the oceans of Earth, for example. The presence of oxygen, in the form of ozone, could be a clear indicator of biological activity. Whether these fingerprints of life are detected in a large number of planets or in hardly any planets at all, the results will still be significant in understanding the origin and development of life on Earth. If, for example, the chemical signs of life are found to be common, then it is also likely that these distant worlds will present a spread of ages in the development of life. It would then be possible to distinguish a variety of biochemical signatures that betray how early life changes the planets over billions of years. In effect, the discovery of a population of Earth-like planets would enable us to study the progress of evolution on our own planet.

Darwin could be launched as early as the second decade of this century and placed into its remote orbit after an initial three-year space flight. It may even turn out that this ESA project is merged with a similar mission being considered by NASA, called the Terrestrial Planet Finder. Besides discovering other worlds like the Earth, the technologies developed from these missions would pave the way for even more ambitious interferometric designs in space. These designs could involve the connection of signals between individual telescopes placed thousands of kilometres apart.

Looking beyond light

While a substantial proportion of our knowledge of the universe is based on light we receive through telescopes, there are other windows on space waiting to be opened by new technology. One of the most important new portals is the direct detection of gravitational waves arising from stars and galaxies. In forming his theory of general relativity, Albert Einstein predicted that, as massive objects warp space around them, any movement of these bodies should then create disturbances. These disturbances are predicted to propagate as waves moving through space-time, much like the way ripples are created when a stone is thrown into a pond. Ripples in space are called "gravitational waves" and their detection represents a fantastic future area of observational astronomy. Just as light can provide information about the temperature and composition of an object, gravitational waves carry information about its mass and motion. Powerful gravitational waves can be expected from violent and high-energy events in space, such as the mergers of neutron stars and white dwarfs, the aftermath of supernovae explosions, and the motion of black holes at the hearts of colliding galaxies. Studying them would enable us to learn a great deal more about the action of gravity in the universe.

These waves from ripples in space-time are exceedingly weak and in fact have not yet been detected, thus prompting the construction of sophisticated new detectors. In a project jointly sponsored by NASA and ESA, the first dedicated gravitational wave detector to be placed in space is now being designed. The Laser Interferometer Space Antenna (LISA) will search for

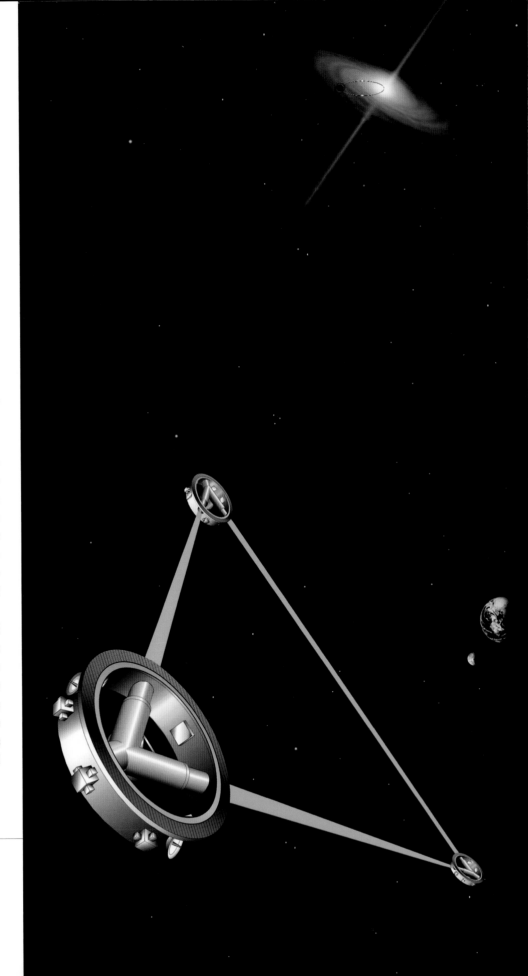

gravitational waves and uniquely probe the universe the way no "normal" telescope can. This exciting new mission will consist of three spacecraft placed into an equilateral triangle, with each apex separated by five million kilometres (three million miles). The positions of the spacecraft will be very accurately tracked to measure the effects of passing gravitational waves, which act to create distortions and alter the relative positions of points in space. Each of the identical spacecraft will contain at its centre a tiny box containing free floating cubes or "test masses". The cubes will be completely shielded from other external disturbances, and it is their minuscule relative movement, accurately measured using laser technology, that will betray the action of gravitational waves

The Laser Interferometer Space Antenna could be launched by 2011, and placed 50 million km (30 million miles) from Earth in order to reduce the variable effects of Earth's gravity. During a planned five-year lifespan, it will detect gravitational wave sources from a variety of directions in the sky. Beside testing Einstein's theories, studies of gravitational waves will provide unique information about the interiors of extreme-gravity regions of space, and improve our understanding of the basic physical laws governing the universe.

The proposed triangular arrangement of the LISA observatory in space. This innovative and challenging structure is designed to open up new windows on the universe by detecting gravitational waves from distant stars and galaxies.

Sailing to the stars

Over the past 40 years our exploration of the universe has been based not just on greater telescope power, but also on fascinating travels through our solar system that have seen spacecraft visit all of the planets except Pluto. With continued conviction and international cooperation, the next couple of decades could witness more ambitious and dramatic missions, ranging from robotic stations on Earth's Moon and drilling into the sub-surface oceans of Europa, to perhaps even a manned mission to Mars. There is no doubt that these are all complex undertakings that will demand new and improved technologies, but they will also focus the ingenuity of future generations. The realization of more ambitious direct exploration of our solar system will subsequently bring closer an even more dramatic achievement for humanity, namely the launch of spacecraft out of our solar system and toward the nearest stars. Our final look ahead in this book is to the remarkable possibility of using solar sails as a method of propelling spacecraft over great distances.

The problem with the current propulsion technology used in spacecraft is that it is heavily propellant-based, with booster rockets applying short powerful thrusts over the first part of a journey, after which the craft coasts to its destination. This method is not feasible for great distances as progress is slow and the amount of on-board fuel required would be enormous. Technological developments over the past few years have given rise to the possibility of future spacecraft being propellant-free, driven instead by solar sails. This exciting innovation would utilize the radiation from the Sun as its principal source of propulsion. Following launch from Earth, a vast lightweight sail spanning several hundred metres would be unfurled to "trap" solar radiation; momentum would be provided by photons of light exerting pressure on the sail, much like the wind sails of a ship. It is anticipated that solar sails would eventually enable travel at speeds five times greater than the rocket propulsion systems of today. Performance of the sails could even be enhanced by the use of high-powered lasers for brief periods as a supplementary energy source. Indeed, microwave and laser driven sails have already been tested in laboratories by NASA scientists.

The sails would have to be constructed from highly reflective lightweight material capable of resisting electrical charge and extreme temperatures. The technology for making them is within reach, exploiting for example tough carbon-fibre material, coated on the Sun-facing side with a thin layer of reflective aluminium. Like a giant mirror, such a sail would reflect sunlight and, using the momentum of photons, propel the craft through space. A greater engineering challenge is likely to come from working out how to package, deploy, and operate the sail in space.

The development of giant new telescopes will certainly expand our visions of the universe over the next two decades, but the greatest quest would be the deployment of spacecraft capable of accelerating to the nearest stars. This technological feat will then transform our understanding of stellar environments, similar to the manner in which space missions have advanced our knowledge of the solar system.

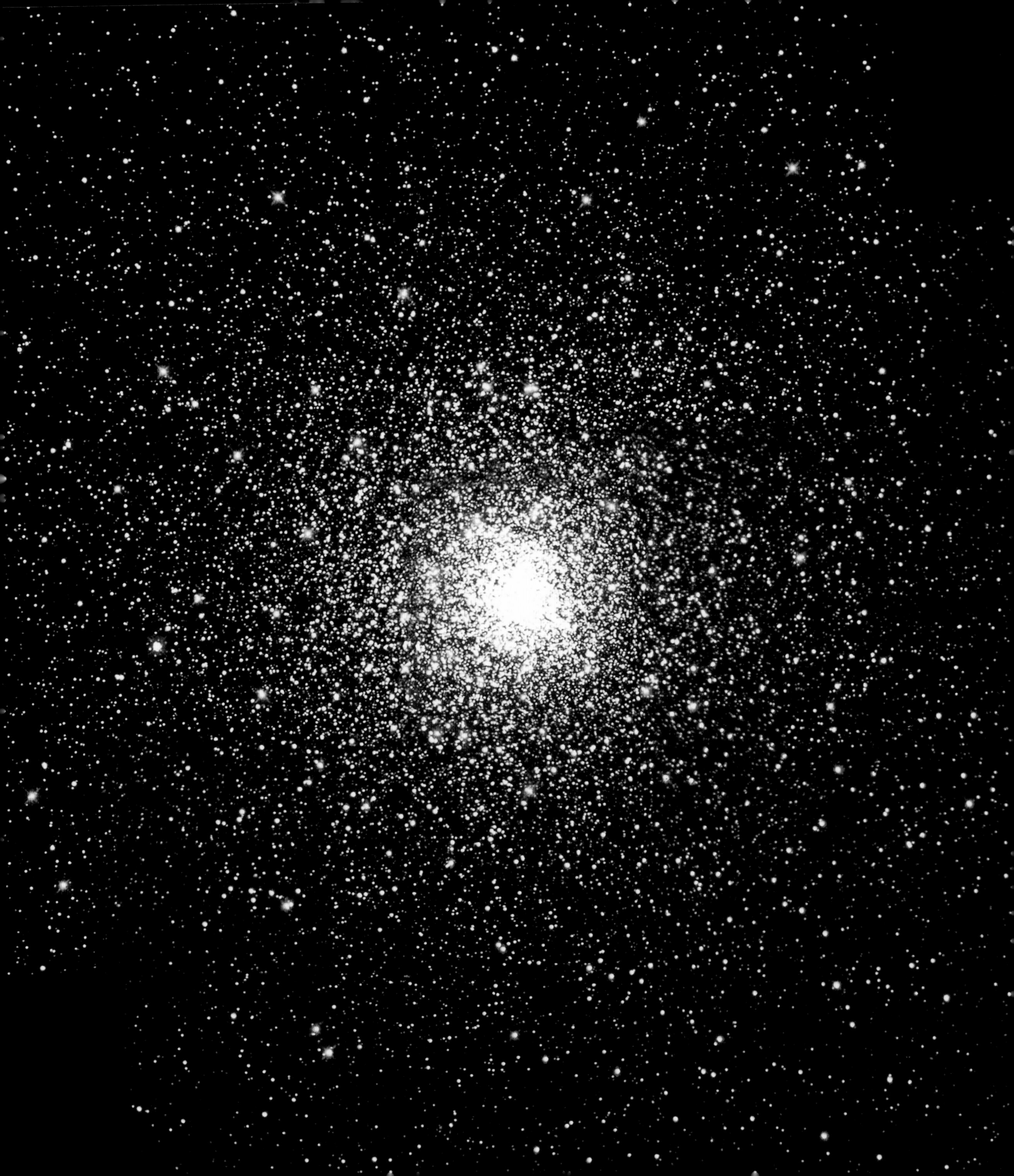

This sequence shows the faint first detection from telescopes on Earth of the Pluto-like planetoid named Sedna (arrowed). This object is three times further away from Earth than Pluto, making it the most distant currently known body in the solar system. Future spacecraft may extensively explore these deep outer regions, with the expectation of finding many more planetoids like Sedna.

The current vision is that an interstellar probe driven by solar (and laser) sails could be launched by the end of the second decade of this century. A sail nearly half a kilometre (1640ft) across would be deployed in space, which, after swinging close to the Sun for an initial intense push from solar radiation, would gradually accelerate over weeks and months. Eventually travelling at 320,000 million km per hour (200,000 miles per hour), the mission would cover in just eight years the same distance that *Voyager 1* has travelled in more than four decades, since it explored Jupiter and Saturn in the late 1970s. Such an interstellar probe could provide amazing new insights into the outermost regions of our solar system, then beyond into the medium between the stars. The goal is that the craft would arrive in a region at least 200 times further from the Sun than the Earth is, after cruising for only a few years. As it travelled further beyond Pluto, a great deal would be revealed about the history of the solar system. Of specific interest is the first detailed exploration of the Oort cloud, which is a vast reservoir of icy comets that extends almost halfway to the nearest star. An example of the kind of object awaiting discovery in the Oort cloud is the recently uncovered Pluto-like planetoid called Sedna. At its furthest, Sedna is 130 billion km (80 billion miles) from the Sun, which is nearly 900 times the distance between the Earth and the Sun. Sedna is the most distant solar system object discovered so far, appearing only as a faint point of light through powerful telescopes on Earth. Sedna is less than 1700km (1000 miles) across, and it is anticipated that a space-craft mission travelling through the Oort cloud would encounter many more objects like Sedna.

Aside from the discoveries the probe makes about our solar system and its neighbourhood, it will also serve as a precursor to a wider range of future spacecraft. There will be a critical need to demonstrate the success and feasibility of solar sails as a reliable, high-speed propulsion method. The immediate reward would be to ease the delivery of payloads to future outposts on the Moon and Mars. Beyond that lies the daring goal of reaching the nearest stars, starting with the Alpha Centaurus star system at a distance of almost four and half light years. A tremendous variety of technological breakthroughs will have to be achieved before such an ambitious plan can be realized, not least making the craft sufficiently autonomous and developing a method of communicating over the vast distances involved. Exotic propulsion ideas speculated about at research centres today range from powerful sails to the roles of antimatter, modified gravity, and quantum teleportation.

The next two decades could define new frontiers in human exploration, while simultaneously raising complex scientific and philosophical issues. Ambitious space projects, with well-defined objectives, offer the most exciting complement to the herculean telescopes of the next generation. Together, these great advances stand to broaden, beyond imagination, our visions of the universe.

Lightweight sails powered by the radiation from the Sun and intense space lasers may allow future spacecraft to accelerate to enormous speeds, therefore making flights to the outermost solar system and beyond feasible and practical.

Glossary

accretion gradual growth in mass, such as when a star forms by steadily accumulating gas.

accretion disc disc of matter spiralling in toward a massive object such as a star or black hole.

angular diameter angle subtended by the diameter of an object as seen at a distance.

asteroid small, rocky body in space.

aurora light display that results when charged particles from the solar wind enter the Earth's atmosphere around the poles: sometimes called Northern or Southern Lights.

Big Bang theory theory of cosmology to describe the expansion of the universe, which is presumed to have begun with an explosive event.

binary star two stars that are gravitationally bound together and orbit one another. Many stars are found to be in binary-star systems.

black hole object or region of space with such strong gravitational pull that nothing (not even light) can escape.

brown dwarf larger than planets, but not massive enough to undergo nuclear fusion of hydrogen in its core and become a star; brown dwarves are "failed stars".

Cepheid variable class of luminous pulsating stars with associated cyclic changes in brightness. The period of repetition is related to the true luminosity of the star, making them useful for measuring astronomical distances.

chromosphere lower atmosphere of the Sun, lying between the photosphere and the corona.

comet small body made mostly of dust and ice, in an elliptical orbit about the Sun. As a comet nears the Sun, some of its material boils off to form a long tail.

compound substance that can be broken down into elements.

constellation grouping of stars identified on the celestial sphere, often named after characters or beasts from ancient mythology.

corona tenuous and hot outer atmosphere of the Sun.

coronagraph instrument used to block the central light from a star so the surrounding region can be viewed.

coronal mass ejection transient event in which billions of tonnes of gas is blasted away from the Sun into space.

cosmic microwave background radiation (CMBR) relic glow related to the Big Bang origin of the universe. The CMBR is uniformly detected as microwave radiation coming from all directions in space.

cosmic rays particles such as electrons and protons that travel through space close to the speed of light.

cosmology study of the structure and evolution of the universe as a whole.

crust outer surface layer of a terrestrial planet.

dark energy energy that could be causing the accelerated expansion of the universe.

dark matter enigmatic component of the universe with the attributes of mass and therefore gravity, but giving off no light capable of being detected on the electromagnetic spectrum.

density amount of mass contained in a certain volume.

disc flattened rotating structure of gas and possibly dust.

Doppler effect change in colour of light when the source of the light and the instrument of observation are moving with respect to one another. A similar effect can be noted in the pitch of a whistle as a train passes by.

dust microscopic bits of solid matter found in interstellar or interplanetary space.

eclipse event in space during which one object passes in front of another.

electromagnetic radiation or **light** consisting of waves moving through the electric and magnetic fields.

electromagnetic spectrum complete range of light as characterized by wavelength, frequency, or energy. Examples include X-ray light, ultraviolet light, and visible light.

electron negatively charged subatomic particle that normally moves about the nucleus of an atom.

element substance that cannot be reduced by chemical processes into a simpler substance.

elliptical galaxy galaxy that appears oval-shaped. Elliptical galaxies lack substantial amounts of interstellar gas and are mostly comprised of older and redder stars.

erosion wearing down of geological features due to the action of water, wind, ice, and so on.

event horizon boundary that marks the point of no return between a black hole and the outside universe.

exo-biology study of life as it might occur elsewhere.

extra-solar planet or **exo-planet** planet orbiting around a star other than the Sun.

extra-galactic anything beyond our Milky Way Galaxy.

flare explosive event occurring in or near an active region on the Sun.

fusion (nuclear) process by which light atomic nuclei are combined into heavier ones, with a release of energy.

galactic cannibalism merger in which a larger galaxy consumes a smaller one.

galaxy collections of millions and billions of gravitationally bound stars, plus gas and dust

gamma rays most energetic form of electromagnetic radiation.

general relativity (theory of) theory developed by Albert Einstein in the early 1900s to relate gravity, acceleration, and the structure of space.

giant (star) evolved phase of a star's life when the core of a star has exhausted its supply of hydrogen for nuclear fusion. As a result, the star may swell to between ten and 100 times the size of the Sun.

globules (Bok) small, round and dark cloud of gas and dust in interstellar space.

globular cluster collection of many thousands, and sometimes millions, of stars that are gravitationally bound and located in the halos of galaxies.

granules (on Sun) circulating loops of gas seen in the surface layer of the Sun called the photosphere.

gravity mutual attraction between objects with mass; the greater the mass, the stronger its gravitational pull.

gravitational lens chance alignment of objects in space, one of which provides multiple images of the other by gravitationally bending its light.

gravitational waves it is predicted by the theory of general relativity that gravitational waves propagate at the speed of light and create distortions in space.

halo (of galaxy) the outermost regions of a galaxy, which contain a spherical, sparse distribution of stars and clusters.

heavy elements elements other than hydrogen or helium.

helioseismology study of the vibrations of the Sun.

Hubble law law that relates to the observed velocity of recession of a galaxy to its distance from us.

hypernova a scaled-up version of a supernova explosion, involving a very massive star.

inflation (universe) hypothetical phase in the very early universe when space is thought to have expanded extremely rapidly for a short time.

infrared radiation part of the electromagnetic spectrum just beyond the red end of the visible range.

interstellar medium gas and dust intermixed in the space between stars.

irregular galaxy galaxy without obvious symmetry and not adhering to the classes of spiral or elliptical galaxies.

jovian planet (giant planet) large planet resembling Jupiter, that is mostly composed of gas, not rock or metal.

large-scale structure (of the universe) structure of the universe on the scales of several hundred million light years, involving clusters of galaxies.

laser device for amplifying a light signal at a particular wavelength into a coherent beam.

light year the distance travelled by light in a vacuum in one year. One light year is equal to 9460 billion km (5880 billion miles).

Local Group a collection of galaxies including the Milky Way Galaxy and its nearest neighbours.

magnetic fields region of space associated with a magnetized object within which magnetic forces are detected.

magnetic poles points on a magnetic body (such as the Earth) at which the density of magnetic lines is greatest.

mantle (of Earth) a layer of the Earth's interior that lies above the core and just below the crust.

mass a measure of the amount of matter within a body.

merger (of galaxies) the result of two galaxies colliding to form a new galaxy.

Messier (catalogue names, such as M31) catalogue of nebulae, star clusters, and galaxies compiled by Charles Messier in the late 18th century.

metals an element other than hydrogen or helium. In the context of planets, the term refers to substances that are good conductors of electricity such as iron, tin, and so on.

meteorite remains from a meteor that survives passage through the atmosphere and strikes the ground.

microorganism or **microbe** an organism that is so small that it can be seen only with a microscope.

Milky Way band of light that encircles the night sky, due to the stars and nebulae lying close to the disc of our Galaxy.

molecular cloud cold, dense interstellar cloud of mostly hydrogen molecules.

momentum measure of the state of motion of a body, defined as the product of its mass and velocity.

nebula cloud of gas and dust in space.

neutrino a particle that travels at or near the speed of light, having no charge and little or no mass.

neutron a subatomic particle with mass similar to a proton but no charge.

neutron star a star of extremely high density made almost entirely of neutrons.

nova a "new star" (usually a white dwarf) that greatly brightens for a time as a result of an explosive event.

nuclear referring to the nucleus of an atom.

nucleus the heavy central part of an atom consisting of protons and neutrons, which is orbited by one or more electrons; the central region of a galaxy.

orbit path of astronomical body about another body or point.

outflow flow of gas and particles away from a star or galaxy.

outgassing the release of gases into a planet's atmosphere due to volcanic activity.

ozone layer of gas in the atmosphere on an Earth-like planet at a height of tens of kilometres above the surface, where incoming ultraviolet radiation from the Sun is absorbed.

photon discrete amount of light energy.

photosphere region of the Sun's atmosphere from which visible light escapes into space.

planet large body (rocky or gaseous) that orbits a star.

planetary nebula ejected outer layers of a red giant star, spread over a volume the size of our entire solar system.

plasma gas that is fully or partially ionized.

pressure force spread over a given area.

progenitor star eventually explodes as a supernova.

prominence loop or sheet of glowing gas ejected from an active region above the Sun's photosphere.

proton subatomic particle with charge opposite to an electron, but having a much greater mass.

proto-planet a planet in the process of forming.

proto-star collapsing mass of gas and dust out of which a star will be born.

pulsar spinning neutron star with strong magnetic fields that accelerate and eject high-energy particles. The ejected radiation is detected as short, regular radio pulses at the Earth.

quasar nucleus of a very active galaxy which appears star-like, but is at a great extra-galactic distance from us.

radar technique of bouncing radio waves off an object, then detecting the radiation that the object reflects back.

radiation refers to electromagnetic radiation, such as visible light, infrared, ultraviolet and so on.

radio telescope designed to collect and detect radiation at radio wavelengths.

red giant a large, cool star with a high luminosity, low surface temperature, and large proportions.

redshift the shift to longer wavelengths of the light emitted by distant galaxies that are receding from us.

regolith layer of tiny rock fragments that cover the surface of Earth's Moon.

sediment deposits and cementation of fine grains of material usually resulting from erosion in lakes and oceans.

solar eclipse an eclipse of the Sun by the Moon, caused by the passage of the Moon in front of the Sun.

solar wind stream of charged particles that escape from the Sun's atmosphere at high speeds and flows out into the solar system.

solvent liquid capable of dissolving or dispersing another substance.

spectral line radiation at a particular wavelength of light produced by the emission or absorption of energy by an atom.

spectrograph instrument attached to a telescope to create an image of a spectrum.

spiral arms structures containing young stars and interstellar material that wind out from the central regions of some galaxies.

spiral galaxy a galaxy where most of the gas and stars are in a flattened disc that displays spiral arm structures. Our Milky Way Galaxy is a spiral galaxy.

star a massive sphere of gas that shines by generating its own power.

starburst a galaxy, or region within a galaxy, that is experiencing a tremendously high rate of star formation.

star cluster stars gravitationally bound to one another.

stellar evolution changes in structure and properties of a star over the course of time, usually over millions of years.

stellar wind the outflow of gaseous material at high speeds from a star.

subatomic particle any particle that is small compared to the size of an atom.

sunspot a slightly cooler temporary region on the Sun's surface that appears dark by contrast to the surrounding hotter regions.

super-cluster a collection of several clusters of galaxies that spans a large region of space, but is not necessarily gravitationally bound.

supergiant an evolved phase of stellar evolution, supergiants are massive stars that can swell up to several thousand times bigger than the Sun.

supernova one of the brightest and most violent events in the universe which occurs when a star explodes. A Type I supernova occurs when a white dwarf accretes matter in a binary system. A Type II supernova results from the implosion of a massive star at the end of its life.

temperature a measure of how hot or cold an object is; a measure of the average random speeds of microscopic particles in a substance.

terrestrial planet a planet that is predominantly composed of rocky and metal substances. Earth, Venus, and Mars are terrestrial planets.

thermonuclear energy results from particle encounters that are given high velocities through heating.

tidal force differences in the force of gravity across a body that is being attracted by another larger body. The result may be the deformation of the smaller body.

ultraviolet radiation electromagnetic radiation of the region just outside the visible range, corresponding to wavelengths slightly shorter than blue light.

universe the total of all space, time, matter, and energy.

vent a hole or opening in the crust of a planet.

velocity rate and direction in which distance is covered over some interval of time.

void regions of space between super-clusters of galaxies that appears to lack luminous matter.

volcano eruption of hot lava from below a planet or moon's crust to the surface.

waveband a broad region of the electromagnetic spectrum, such as ultraviolet, X-ray, and optical.

white dwarf the collapsed end-state of a star that has exhausted its nuclear fuel, and shines from residual heat.

WIMPs (weakly interacting massive particles) a hypothetical particle that may constitute part of the unseen dark matter in the universe.

X-rays high-energy electromagnetic radiation, with photons of wavelengths intermediate between those of ultraviolet radiation and gamma rays.

Index

Visions of the Universe

Acknowledgements

PHOTOGRAPHIC ACKNOWLEDGEMENTS

Key: ACS Advanced Camera for Surveys; APOD Astronomy Picture of the Day; ASU Arizona State University; AURA Association of Universities for Research in Astronomy; Caltech California Institute of Technology; CXC Chandra X-Ray Center; DLR Deutsches Forschungsanstalt für Luft und Raumfahrt; ESA European Space Agency; ESO European Southern Observatory; FU Freie Universität; GSFC Goddard Space Flight Center; IAEF Institut für Astrophysik und Extraterrestrische Forschung der Universität Bonn; IPAC Infrared Processing and Analysis Center; JHU John Hopkins University; JPL Jet Propulsion Laboratory; MIT Massachusetts Institute of Technology; MLO Mount Laguna Observatory; MSSS Malin Space Science Systems; NAOJ National Astronomical Observatory of Japan; NASA National Aeronautics and Space Administration; NOAO National Optical Astronomy Observatory; NRAO National Radio Astronomy Observatory; NRL Naval Research Laboratory; NSF National Science Foundation; PIRL Planetary Image Research Laboratory; PPARC Particle Physics and Astronomy Research Council; SAO Smithsonian Astrophysics Observatory; SDSU South Dakota State University; SNO Sudbury Neutrino Observatory; STScI Space Telescope Science Institute; UCO/LO University of California Observatories/Llick Observatory; UCSB University of California, Santa Barbara; UIUC University of Illinois at Urbana-Champaign; 2MASS Two Micron All-Sky Survey; VLA Very Large Array; WMAP Wilkinson Microwave Anisotropy Probe

2 NASA/Hubble Heritage Team, STScI; **5** as 32; **7** courtesy APOD/M Schirmer, IAEF/W Gieren, University de Concepcion, Chile/ESO; **8–9** Science Photo Library/STScI/NASA; **11** ESA/DLR/G Neukum, FU Berlin; **12–13** NASA/JPL/Cornell; **14** NASA/GSFC/image by Reto Stöckli, enhancements by Robert Simmon; **16–17** NASA/JPL/Cornell; **18–19**, **20**, **21** NASA/JPL/MSSS; **23** Science Photo Library/NASA; **24** NASA/JPL/US Geological Survey; **27** NASA/JPL/DLR; **28** NASA/JPL/University of Arizona; **29** Science Photo Library/NASA; **30** NASA/JPL/PIRL/University of Arizona; **32–33** NASA/JPL/STScI; **35**, **36–7** NASA/JPL/US Geological Survey; **38–39** NASA/JPL; **40–41** NASA/STScI/C R O'Dell and S K Wong, Rice University; **42** NRL; **43** Science Photo Library/NASA; **44** Science Photo Library/STScI/NASA; **46–47** NASA/STScI/G Bacon; **48–49** PPARC/David A Hardy, astroart.org; **52–53** NASA/JPL/NOAO/ESA/STScI and the Hubble Heritage Team; **54** Science Photo Library/STScI/NASA; **56–57** NASA/STScI/ESA/M Romaniello, ESO; **58** NASA/STScI/ESA/J Hester, ASU; **60–61** NASA/M Clampin, G Hartig, STScI/H Ford, JHU/G Illingworth, UCO/LO/ASC Science Team/ESA; **62-3** ESO; **64–65** Science Photo Library/Mark Garlick; **67** NASA/STScI; **68-9** Science Photo Library/STScI/NASA; **70–71** NASA/STScI/ESA/R Thompson, University of Arizona; **72–73**, **75** Science Photo Library/STScI/NASA; **77**, **78–79** NASA/GSFC/WMAP Science Team; **82–83** X-ray: NASA/UIUC/Y Chu & R Gruendel, optical SDSU/MLO/Y Chu et al; **84** NASA/A Fujii, D Malin, David Malin Images/STScI/Hubble Heritage Team; **87** NASA; **88–89** NASA/GSFC; **90–91** Science Photo Library/NASA; **92–93** Royal Swedish Academy of Sciences/Institute for Solar Physics; **95–96** NASA; **97** X-ray: NASA/CXC/Rutgers/J Warren et al, optical NASA/STScI/UIUC/Y Chu; **98–99** Science Photo Library/STScI; **100–101** NASA/H E Bond, STScI/ESA; **102–103** NASA/NOAO/ESA/the Hubble Helix Nebula Team/M Meixner, STScI/T A Rector, NRAO; **104–105** NASA/CXC/J Hester et al, ASU; **106–107** ASA/JPL/STScI, Hubble Heritage Team; **109** Subaru Telescope/NAOJ. All rights reserved; **110–111** NASA/STScI, Hubble Heritage Team; **112–113** NASA/M Clampin & G Hartig, STScI/H Ford, JHU/G Illingworth, UCSC/LO/ACS Science Team, ESA; **114** NASA/STScI, Hubble Heritage Team; **115** Subaru Telescope, NAOJ. All rights reserved; **116–117**, **118–9** NASA/F Summers, STScI/C Mihos, Case Western Reserve University/L Hernquist, Harvard University; **120–121** NASA/JPL/2MASS; **122–123** NASA/Caltech/IPAC/2 MASS; **124** NASA/Don Figer, STScI; **125** X-ray (blue) NASA/CXC/Northwestern/F Zadeh et al, (green) Nobeyama/M Tsuboi, (red) NRAO/VLA F Zadeh et al; **126–127** NASA/CXC/MIT/F K Baganoff et al; **129** X-ray: NASA/CXC/M Karovska et al, radio 21cm image NRAO/VLA J Van Gorkom/Schminovich et al, radio continuum image NRAO/VLA/J Condon et al, optical Digital Sky Survey, UK Schmidt Telescope/STScI; **130–131** NASA/CXC/MIT/UCSB/P Ogle et al, optical NASA/STScI/A Capetti et al; **132** NASA/CXC/SAO/H Marshall et al; **134–135** Science Photo Library/STScI/NASA; **136–137** NASA/ESA/T M Brown, STScI; **138** NASA/The Hubble Heritage Team/A Riess, STScI; **141** Science Photo Library/STScI; **142–143** NASA/N Benitez, H Ford, JHU/T Broadhurst, Racah Institute of Physics, Hebrew University of Jerusalem/M Clampin, G Hartig, STScI/G Illingworth, UCO/LO/ACS Science Team/ESA; **144** ESA/NASA/Jean-Paul Kneib, Observatoire Midi-Pyrénées/Caltech; **147** SNO/Lawrence Berkeley National Laboratory; **148** John Rowe Animations/Jodrell Bank Observatory, University of Manchester; **150–151** Corbis/STScI/NASA/Ressmeyer; **153** NASA/CXC/SAO/Hubble Space Telescope; **155** Science Photo Library/STScI/NASA; **156–157** NASA/CXC/ESO; **158–159** NASA/2MASS; **161** Science Photo Library/STScI/NASA; **162–163** NASA/CXC/M Weiss; **165** NASA/CXC/SAO; **166–167** ESA; **168** NASA/John Dowman of John Frassanito & Associates; **170–171** NOAO/AURA/NSF; **172–173** NASA; **176** ESO; **179** ESA 2002. Illustration by Medialab; **181** NASA/JPL; **183** Science Photo Library/STScI/NASA; **185** Michael Carroll/The Planetary Society.

Author's Acknowledgements

The author is very grateful to Vivien Antwi and the staff at Mitchell Beazley for their effort and patience.

This book would not have been possible without the superb facilities of NASA and ESA.